ART&DESIGN

ACADEMY GROUP LTD
42 LEINSTER GARDENS, LONDON W2 3AN
TEL: 0171-402 2141 FAX: 0171-723 9540

EDITOR: Nicola Kearton
PRODUCTION EDITOR: Ramona Khambatta
PROJECT EDITOR: Melissa Larner
ART EDITOR: Alex Young
DESIGN: Artmedia

SUBSCRIPTION OFFICES:
UK:JOHN WILEY & SONS LTD
JOURNALS ADMINISTRATION DEPARTMENT
1 OLDLANDS WAY, BOGNOR REGIS
WEST SUSSEX, PO22 9SA, UK
TEL: 01243 843272; FAX: 01243 843232
E-MAIL: cs journals@wiley.co.uk

USA AND CANADA: JOHN WILEY & SONS, INC
JOURNALS ADMINISTRATION DEPARTMENT
695 THIRD AVENUE
NEW YORK, NY 10158, USA
TEL: 212 850 6645 FAX: 212 850 6021
E-MAIL: subinfo@jwiley.com

ALL OTHER COUNTRIES:
VCH VERLAGSGESELLSCHAFT MBH
POSTFACH 101161
69451 WEINHEIM, GERMANY
TEL: 00 49 6201 606 148 FAX: 00 49 6201 606 117

Subscription rates for 1997 (incl p&p): *Annual subscription price*: UK only £68.00, World DM195 for regular subscribers. *Student rate*: UK only £53.00, World DM156 incl postage and handling charges. *Individual issues*: £19.95/DM45.00 (plus £2.30/DM5 for p&p, per issue ordered).
For the USA and Canada: *Art & Design* is published six times per year (Jan/Feb; Mar/Apr; May/Jun; Jul/Aug; Sept/Oct; and Nov/Dec) by Academy Group Ltd, 42 Leinster Gardens, London W2 3AN, England and distributed by John Wiley & Sons, Inc, Journals Administration Department, 695 Third Avenue, New York, NY 10158, USA. Annual subscription price; US $142.00 including postage and handling charges; special student rates available at $105.00, single issue $29.95. Periodicals postage paid at Jamaica, NY 11431. Air freight and mailing in the USA by Publications Expediting Services Inc, 200 Meacham Ave, Elmont, NY 11003. Send address changes to: 'title', c/o Publications Expediting Services Inc, 200 Meacham Ave, Elmont, NY 11003.

Printed in Italy. All prices are subject to change without notice. [ISSN: 0267-3991]
The full text of *Art & Design* is also available in the electronic versions of the Art Index.

2.99
9
Feb

CONTENTS

ART & DESIGN **MAGAZINE**

'Monkeydoodle' 1997 Goldsmiths College, London • Academy Highlights

ART & DESIGN **PROFILE** *No 56*

SCI-FI AESTHETICS

Guest-edited by Dr Rachel Armstrong

2 Dr Rachel Armstrong *What is Sci-Fi Aesthetics?*
6 Simon Costin – Burning
8 Jim Harold *Andres Serrano: The Sea of Possibility*
10 Roberta Graham – Fallen Angel
12 Kathy Acker *The Gift of Disease*
18 Julie Weir *Nick Broomfield's* Fetishes
22 Catherine Williams *Fernando Arias – What is he Bid?*
24 Predrag Pajdic – The Body as Costume
28 Orlan – Carnal Art
32 Elizabeth DYN – Soul Surgery
34 Cindy Jackson – Beauty is Power
36 Nicola Triscott *Jake & Dinos Make Monsters*
37 Jake Chapman *Fuckface*
40 A Hans Scheirl *Dandy Dust*
44 VNS Matrix *Bitch Mutant Manifesto*
46 Peter Gabriel with Real World – Eve
50 Werner Hammerstingl *Twelve Stories*
52 Michael Guida *Computer Power*
58 Julian Blom *Survival Research Laboratories: The Search for Sentient Machines*
60 Julian Blom *Nanotechnology: The Eve of the Diamond Age*
62 Patrice Caire – C.A.I.R.E's Cyberhead
64 Werner Hammerstingl *Stelarc and the Sum of his Parts*
66 Stelarc *Parasite Visions: Alternate, Intimate and Involuntary Experiences*
70 Tamagotchi
71 Tina Keane – Transposition
72 Rob La Frenais *Taking the Alien Exam*
76 Mariko Mori
78 David Bowie – Alien Archetypes and Proteans
79 Davide de Angelis *Digital Frankenstein – Collaborations with David Bowie*
84 Michael Buhler *The Alien Image*

Art & Design *highlights*

NEXT ISSUE

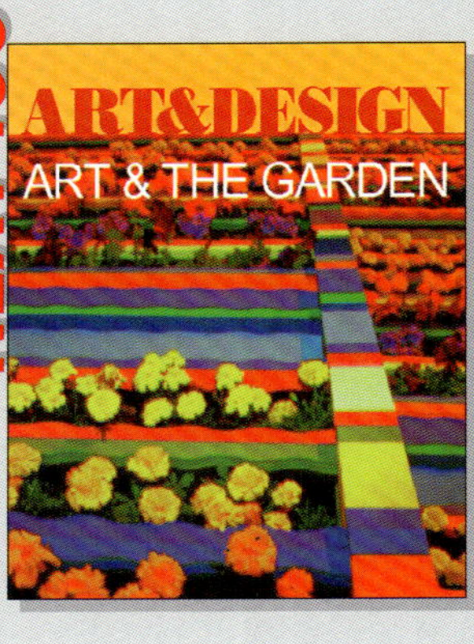

ART & THE GARDEN

Guest-Edited by Anne de Charmant

A&D PROFILE 57

Presenting the garden as a major contemporary art form, this issue of *Art & Design* explores some of the most exciting developments in this field internationally looking at the special place of the garden in our collective consciousness. With contributions from Yves Abrioux on defining the contemporary 'gardenesque'; Stephen Bann exploring the links between land art and the garden landscape; John Stathatos on the continuing battle of Little Sparta – the garden of Ian Hamilton Finlay; Charles Jencks on his application of the New Sciences to his own spectacular garden in Scotland; Günter Nitschke on contemporary Japanese gardens; further essays and features include Ronald Jones' Caesar's Cosmic Garden, Rüdiger Schöttle's Theatre Garden, Derek Jarman's garden at Dungeness, the Festival of Gardening at Chaumont, the designs of Gilles Clément, Bernard Lassus, Newton and Helen Mayer Harrison and Meg Webster.

Anne de Charmant is an art critic and curator based in London, she is also Director of Burford House Gallery in Shropshire.

ISBN: 0-471-97745-4 £19.95 $29.95
305x250 mm, 96 pages
Illustrated throughout, mainly in colour
November 1997

Further information can be obtained from:
(UK only) Academy Group Ltd, 42 Leinster Gardens, London W2 3AN.
Tel: 0171 402 2141 Fax: 0171 723 9540

(USA and Canada) National Book Network, 4720 Boston Way, Lanham, Maryland 20706, USA
Tel: (301) 459 3366 Fax: (301) 459 2118

(Rest of World) VCH, Postfach 101161, 69451 Weinheim, Federal Republic of Germany
Tel: +49 6201 606 144 Fax: +49 6201 606 184

RECENT ISSUES

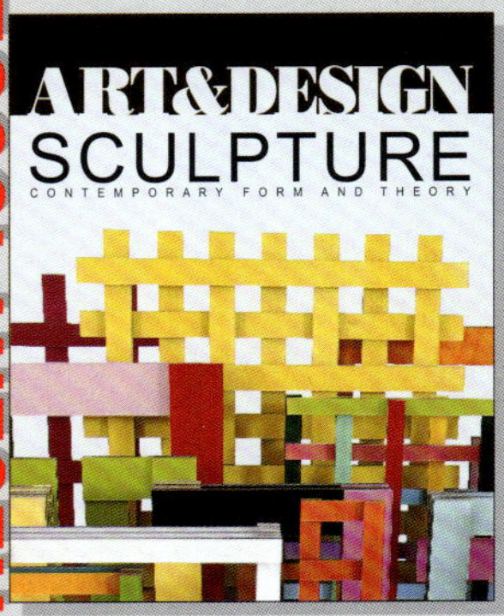

SCULPTURE: CONTEMPORARY FORM AND THEORY

Guest-Edited by Andrew Benjamin

ART & DESIGN PROFILE 55

The renewal of interest in the art object has led to a rethinking of the nature of sculpture. This stimulating issue presents a highly illustrated survey of European and American sculpture from the 1970s to the present day in the context of recent thought. Articles by leading theorists, critics and philosophers include Saul Ostrow looking at postminimal sculpture in the USA; Stan Allen on minimalism in architecture and sculpture; David Batchelor on colour in sculpture; Rebecca Comay on Rachel Whiteread's proposal for a Holocaust Memorial in Vienna; David Moos on Harald Klingelhöller; Charles Harrison on David Batchelor; Jean-Luc Nancy on Lucille Bertrand and Jesse Reiser on Tony Cragg.

Andrew Benjamin is Professor of Philosophy at the University of Warwick and has been widely published, his most recent publications for Academy Editions include *Object • Painting* and *What is Abstraction?*

0-471-97694-6 £19.95 $29.95

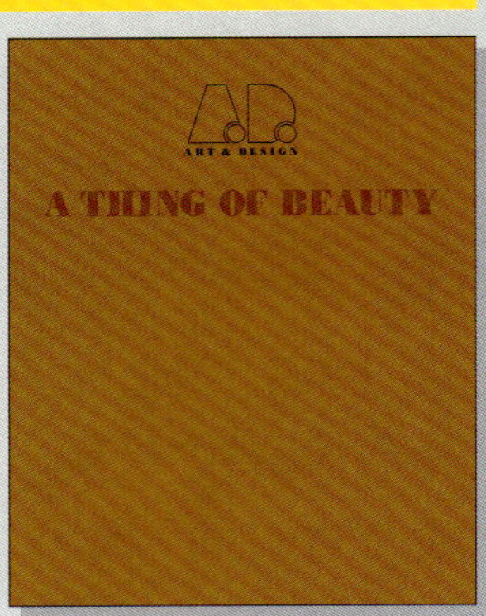

A THING OF BEAUTY IS . . .

Guest-Edited by Michael Petry

A&D PROFILE 54

Perhaps the worst thing that can be said of modern art is that it is beautiful. This dismisses the work as mere fluff, art that is candy, art that is not really art. This profile explores beauty as a cultural construct looking both at its more positive historical associations and how it functions today in relation to artworks in the global media age. With contributions from over 40 different artists and writers, *A Thing of Beauty is . . .* will ask if anything can be allowed to be beautiful again, and if so, what does it mean for art in what is perceived to be an ugly age, or is the myth of the ugly only the other side of the looking glass?

Michael Petry, artist, writer and Co-Director of the Museum of Installation, London, also guest-edited the *A&D* profile *Abstract Eroticism*.

0-471-97684-9 £19.95 $29.95

Academy *highlights*

NEW SCIENCE = NEW ARCHITECTURE?

ARCHITECTURAL DESIGN PROFILE 129

This issue of *Architectural Design* argues that architecture changes only when society experiences fundamental shifts in its basic framework of thought. Centring on the significance of the new science – looking at how far chaos and complexity theories have been accepted as primary explanations – it explores how far architects have chosen to interpret this new paradigm. Including contributions by Charles Jencks, Cecil Balmond, Ben von Berkal. FOA and Kisho Kurokawa, further features include Frank Gehry's Bilbao Museum, Peter Eisenman's Aronoff Center, Enric Miralles' Eurhythmics Centre in Alicante and Daniel Libeskind's V&A proposal.

PB 0-471-97739-X £18.95 $32.50
305x250 mm, 112 pages
Illustrated throughout
September 1997

CHURCH BUILDERS OF THE TWENTIETH CENTURY

Iona Spens and Edwin Heathcote

Tracing developments in church building through the momentous changes in architecture and theology of the 20th century, this book illustrates the astonishing pluralism of modern ecclesiastical architecture which has emerged. Strikingly illustrated, a historical section covering architects such as Auguste Perret, Antoni Gaudí, Frank Lloyd Wright and Le Corbusier leads to a contemporary section featuring the dazzling designs of Tadao Ando, Philip Johnson, Imre Makovecz and Fay Jones as well as many others.

PB 0-471-97755-1 £45.00 $75.00
305x252 mm, 224 pages
Illustrated throughout
September 1997

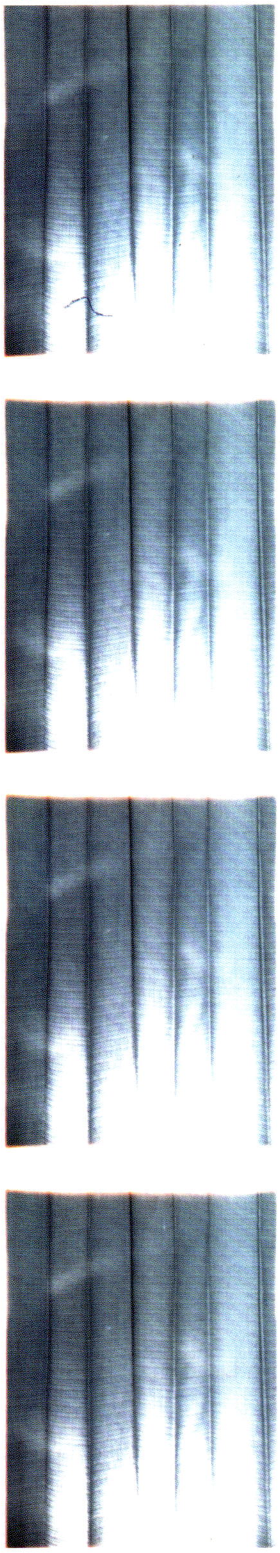

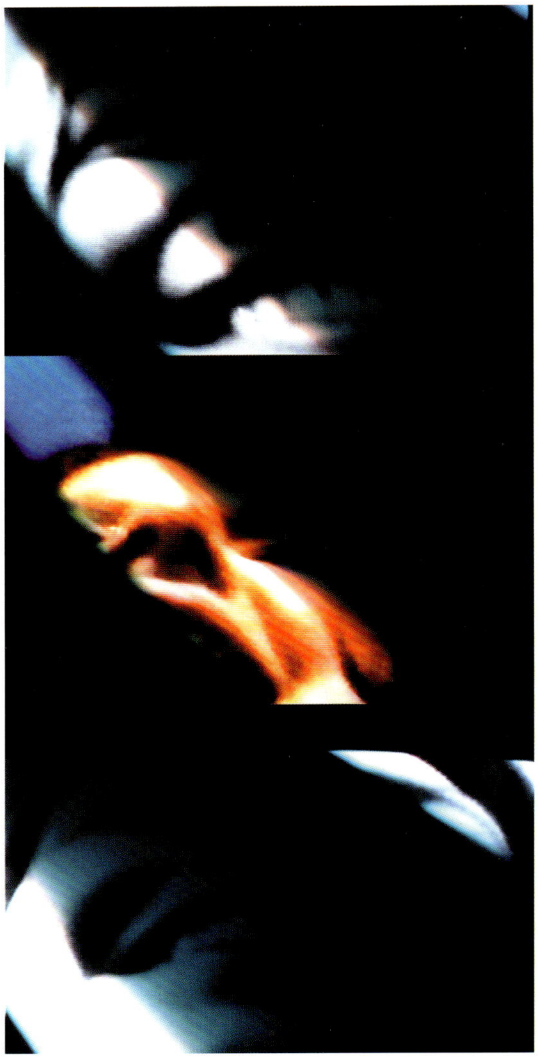

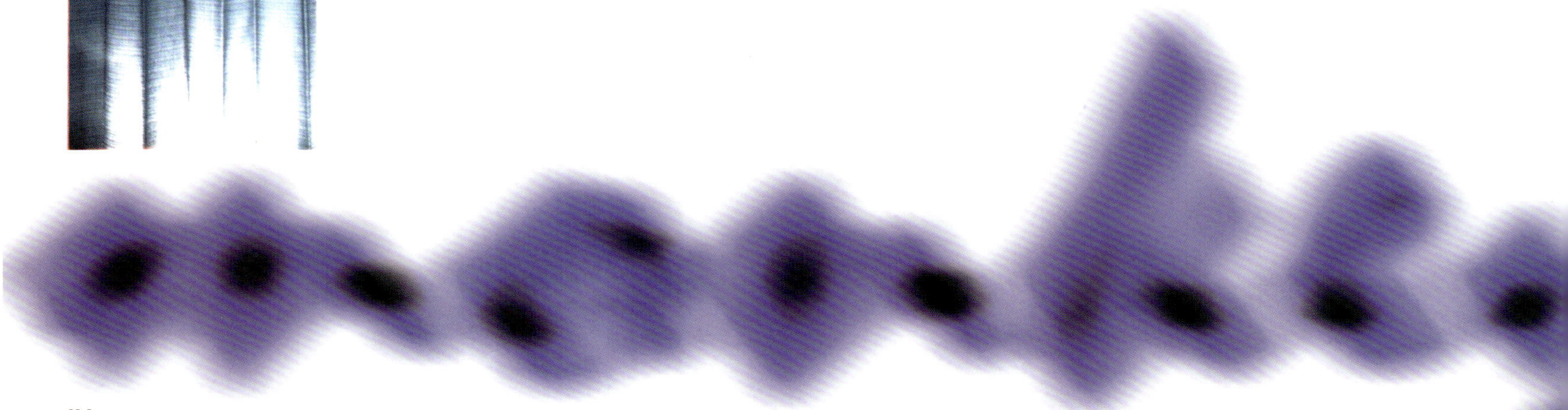

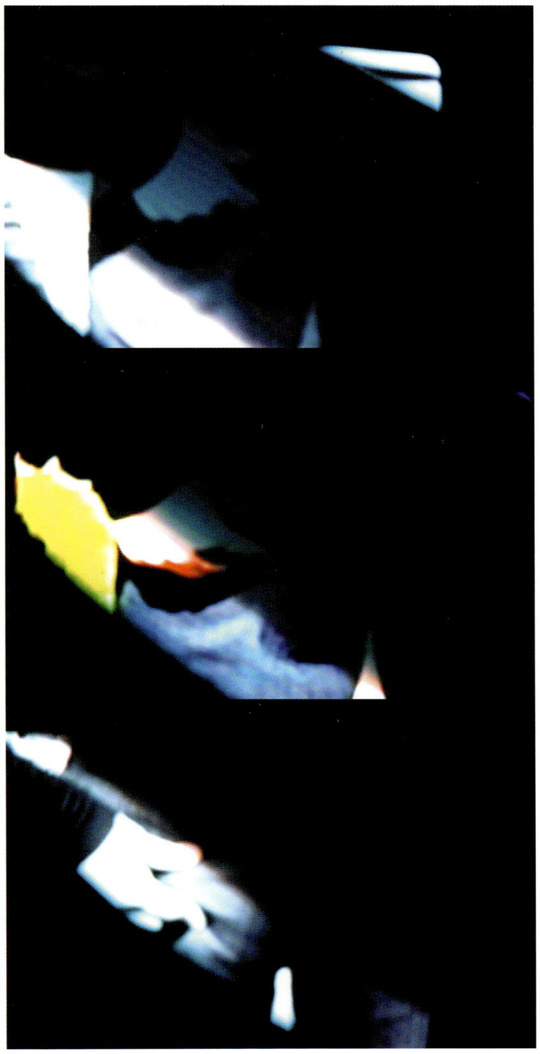

'I have nothing to say, only to show' W Benjamin

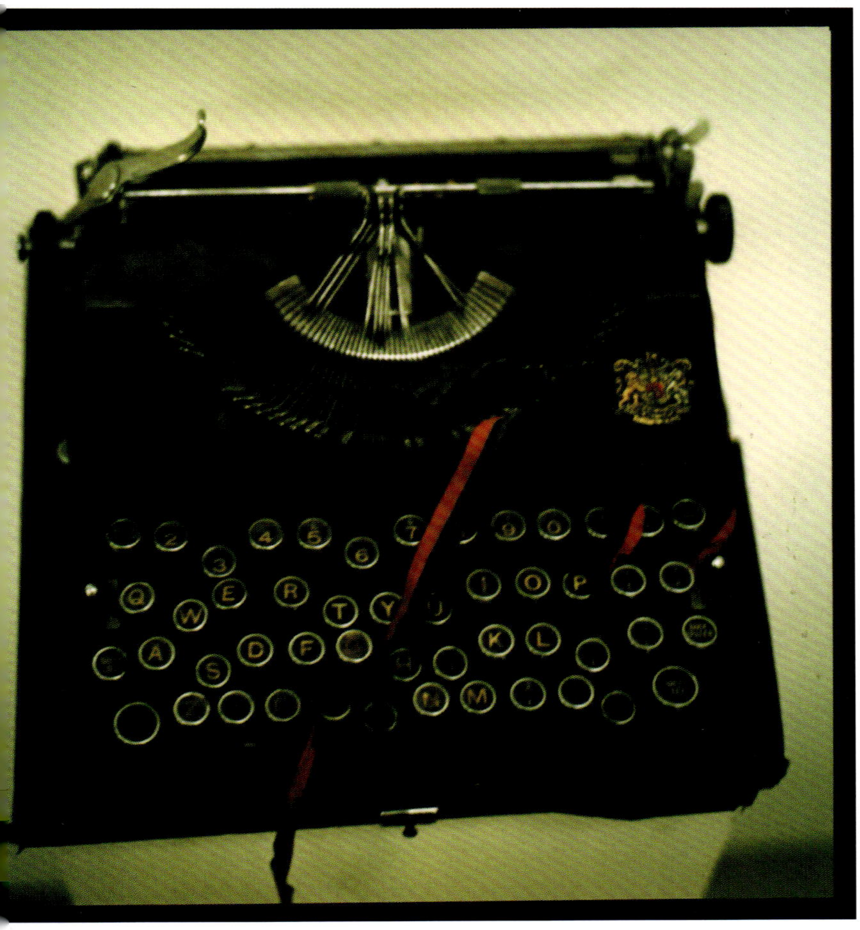

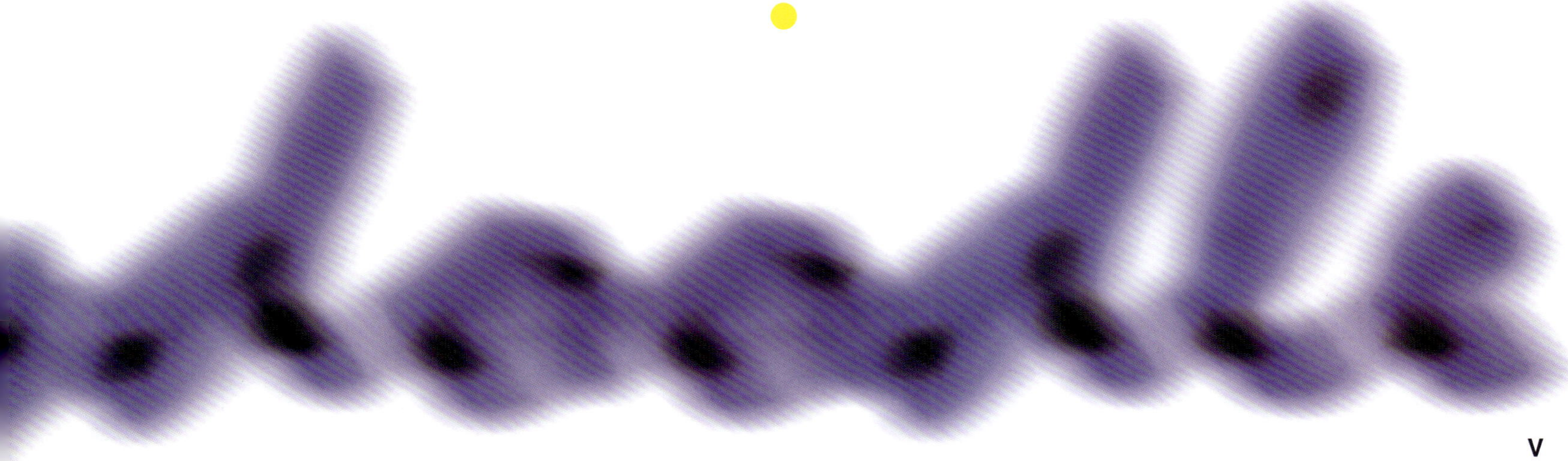

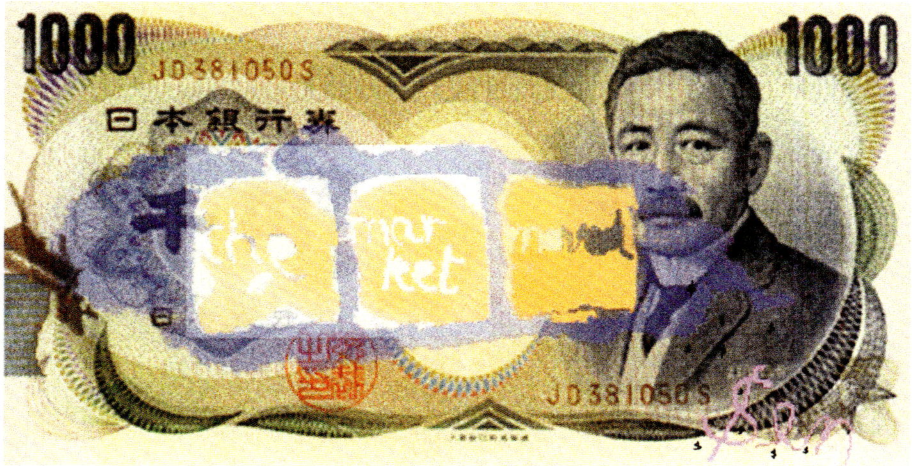

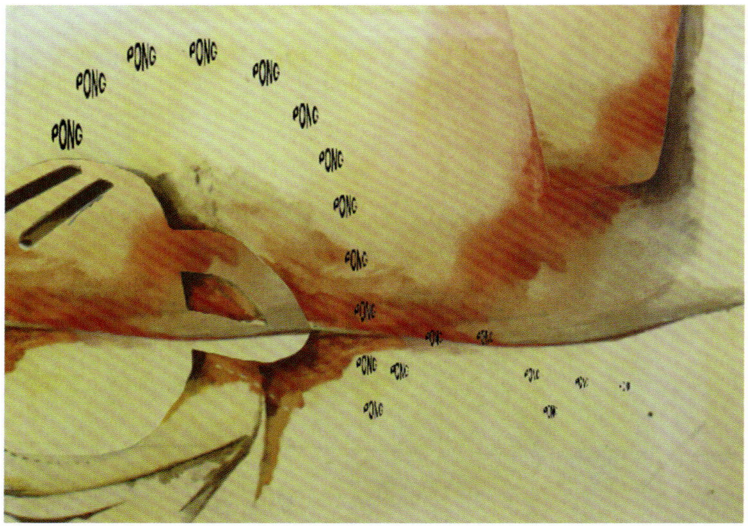

Nancy Solarov
Kyoko Ebata
Ingrid Alexandra McDonald
Stephanie Mulliner
Mélanie Verhille
Soezanna Hildred
Clare Kaufman
Daniel Howard Birt
Claire Waltham-Smith
Tavia Swain
Mary Gordon-Smith

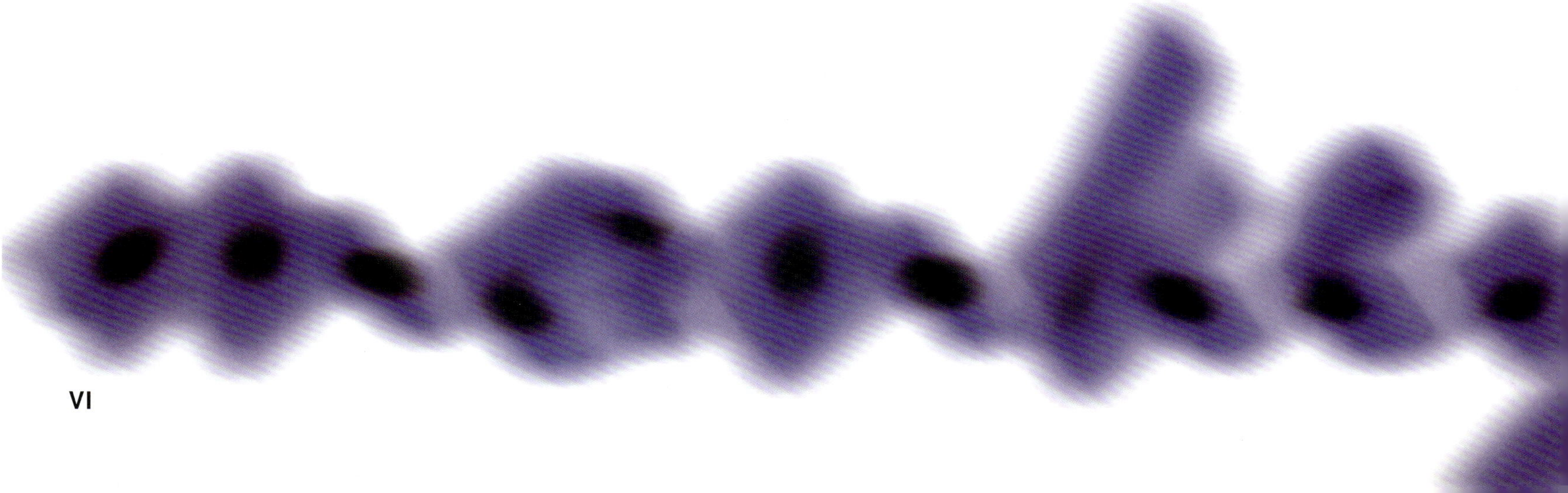

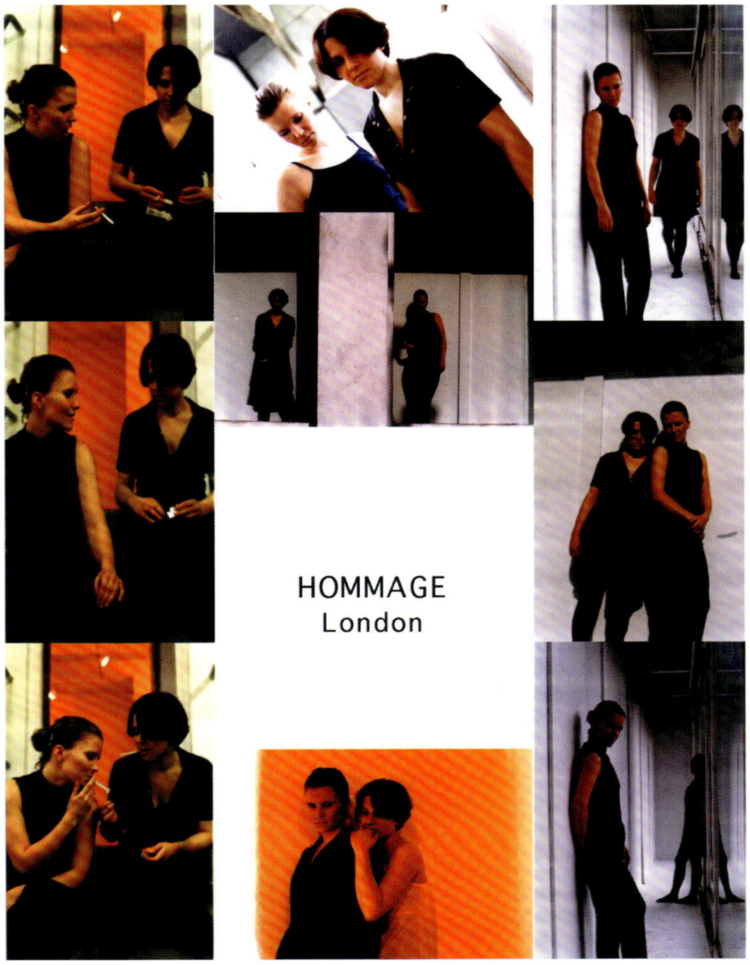
HOMMAGE
London

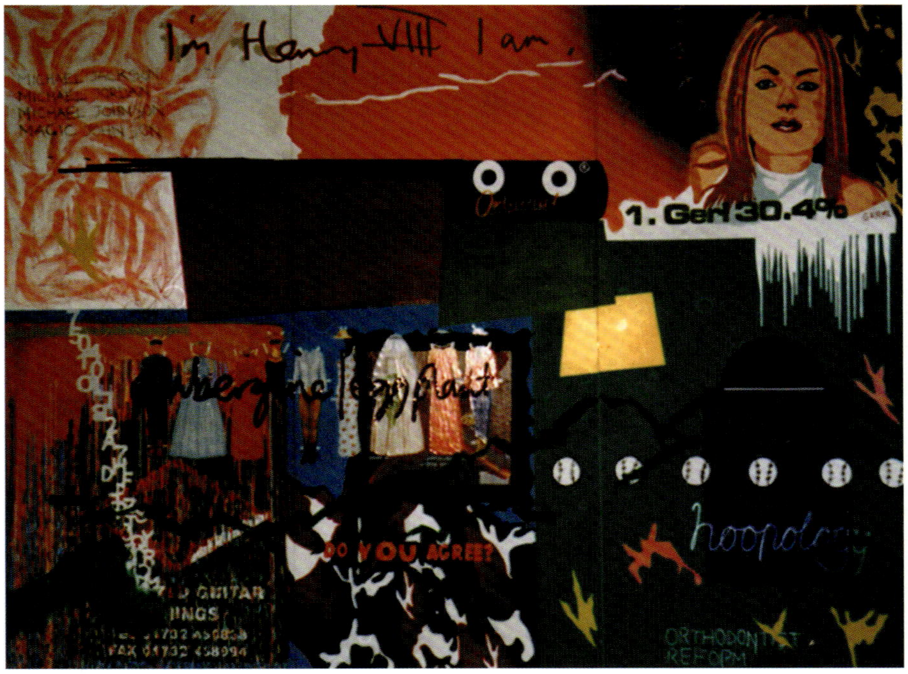
I'm Henry VIII I am
1. Geri 30.4%
DO YOU AGREE?
hoopology
ORTHODONTIST
REFORM

Acknowledgements

We would like to thank Dr Rachel Armstrong for her work in compiling this issue of *Art & Design* along with all the contributors and their galleries for generous contributions. All material is courtesy of the following artists and galleries: *p1:* Paula Cooper Gallery, New York; *pp6-7:* Simon Costin; *pp10-11, p15:* Roberta Graham; *p18:* Nick Broomfield; *p23:* Fernando Arias; *pp24-27:* Predrag Pajdic; *pp28-31:* Orlan; *p32:* Elizabeth DYN; *p34:* Cindy Jackson; *pp36-39:* Victoria Miro Gallery; *pp40-43* A Hans Scheirl; *pp46-49:* Real World, London; *p59:* Ars Futura Gallery, Zurich; *pp62-63:* Patrice Caire; *pp64-65:* This essay was first published on the CD-Rom *Metabody* by Merlin, *p65* images courtesy Stelarc; *pp66-69:* Stelarc; *p70:* Bandai Co Ltd; *p71:* Tina Keane; *pp72-75:* This essay was first published in a longer form in the catalogue for the European Media Festival in Osnabrück, September 1994, and later in the pilot issue of *Mute*, Winter 1994; *pp74-75:* Courtesy of Christopher Bucklow and Laurent Delaye Gallery, London; *pp76-77:* Deitch Projects, New York; *pp78-83:* David Bowie; *pp84-88:* Michael Buhler.

Contributors' Biographies

Dr Rachel Armstrong is a science fiction clinician, qualified doctor, artist and writer based in London. **Jim Harold** is an artist, curator and lecturer. He currently heads the Sculpture programme at Glasgow School of Art and is a contributing editor of *Creative Camera* magazine. **Roberta Graham** is an artist based in London. **Kathy Acker** is the author of *Blood and Guts in Highschool*, *Great Expectations*, *Pussy, King of the Pirates*. She wrote the screen play for Richard Foreman's film *Variety, Birth of a Poet* and most recently, *Requiem*, a libretto for the composer Ken Valitsky which is currently being produced by the American Opera Projects in New York City. Two new books, *Bodies of Work*, a collection of essays, and *Requiem*, a collection of new and old short pieces will be published Autumn 1997. **Julie Weir** is based in the music and multimedia industries and was actively involved with the release and promotion of the film *Fetishes* and its sub-culture. **Catherine Williams** works independently with Fernando Arias and for The Organisation for Visual Arts in London who will tour a major exhibition of Arias' work in the UK during 1998. **Nicola Triscott** is Director of The Arts Catalyst. **Jake Chapman** is an artist based in London. **A Hans Scheirl** is a filmmaker, painter and Dadaist, born in Salzburg, Austria and living in London. He has made 42 neo-aktionist Super-8 films since 1979, he wrote, co-directed and acted in the first cyberdyke feature *Flaming Ears* and directed, wrote and produced the Cybercomicsplatter movie *Dandy Dust*; **VNS Matrix** are a group of Australian artists who work in a variety of mediums including sound, photography, film, performance and multimedia. They formed in 1991 and are Josephine Starrs, Julianne Pierce and Francesca da Rimini. **Werner Hammerstingl** is President of Contemporary Art and Technology Inc, Melbourne. **Michael Guida** is an interactive installation artist and multimedia producer. His work as an artist combines the themes of myth and magic, technology and emotional seduction. Currently he is collaborating on a new work which positions Stanley Milgram's experiments on obedience to authority within the realm of the computer. **Julian Blom** is currently web master at Carlton Television. **Stelarc** is presently a Visiting Fellow at the Studio For Creative Inquiry at Carnegie Mellon University. He has received a Fellowship from the Visual Arts/Craft Board, Australia Council. His artwork is solely represented by Sherman Galleries, Sydney. **Tina Keane** is a multimedia artist and lecturer at the department of film at Central St Martin's College, London. **Rob La Frenais** is a curator and writer concerned with cross-disciplinary issues in art, science and culture. He recently organised the conference and event 'The Incident' in Fribourg, Switzerland and at the ICA in London. He is currently curator for The Arts Catalyst and researcher. **Davide de Angelis** is a multimedia artist and art director. **Michael Buhler** is a painter, art lecturer and UFO researcher. In 1995 he co-produced an animated film *Abductees* for Channel 4 and is currently engaged on other projects involving UFOs and alien imagery.

COVER: Glen Brown, Ornamental Despair (Painting for Ian Curtis) after Chris Foss, *1994, oil, 201x300 cm (Saatchi Collection, London)*
INSIDE COVERS: Brian Eno, generative paintings, 1997 (Opel, London)

EDITOR: Nicola Kearton PRODUCTION EDITOR: Ramona Khambatta
PROJECT EDITOR: Melissa Larner ART EDITOR: Alex Young DESIGN: Artmedia

First published in Great Britain in 1997 by *Art & Design* an imprint of
ACADEMY GROUP LTD

Art & Design Profile 56 is published as part of *Art & Design* Vol 12 9/10 1997
Art & Design Magazine is published six times a year and is available by subscription

Other Wiley Editorial Offices
New York • Weinheim • Brisbane • Singapore • Toronto

Distributed to the trade in the United States of America by
NATIONAL BOOK NETWORK, INC, 4720 BOSTON WAY, LANHAM, MARYLAND 20706

ISBN 0-471-97855-8

Printed and bound in Italy

Art & Design

SCI-FI AESTHETICS

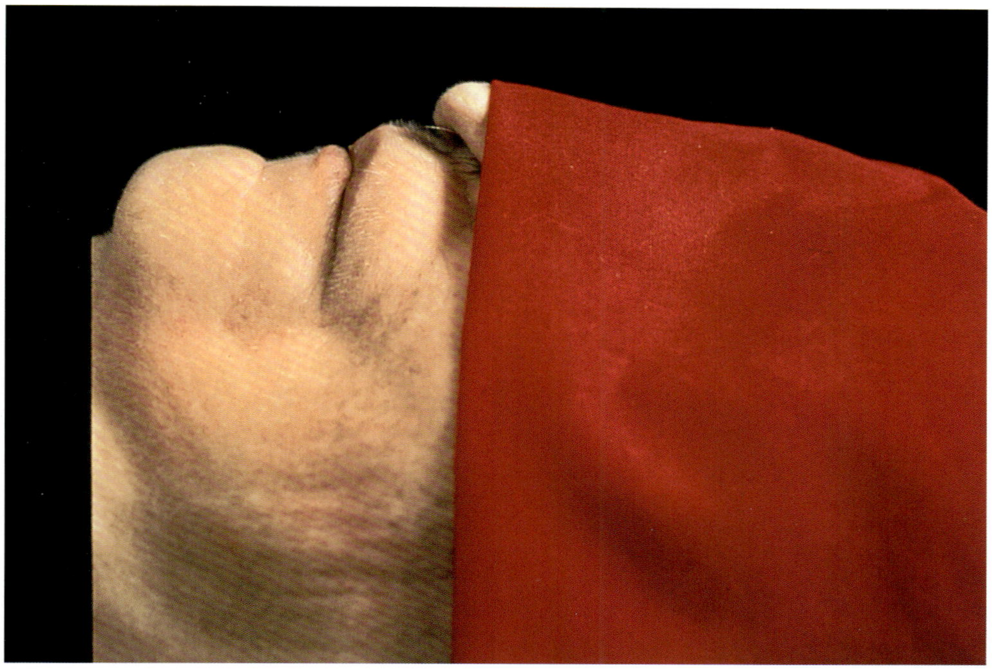

Andres Serrano, *The Morgue (Infectious Pneumonia)*, 1992, Cibachrome, silicone, plexi-glass, wood frame, 125.7 x 152.4cm

ACADEMY EDITIONS • LONDON

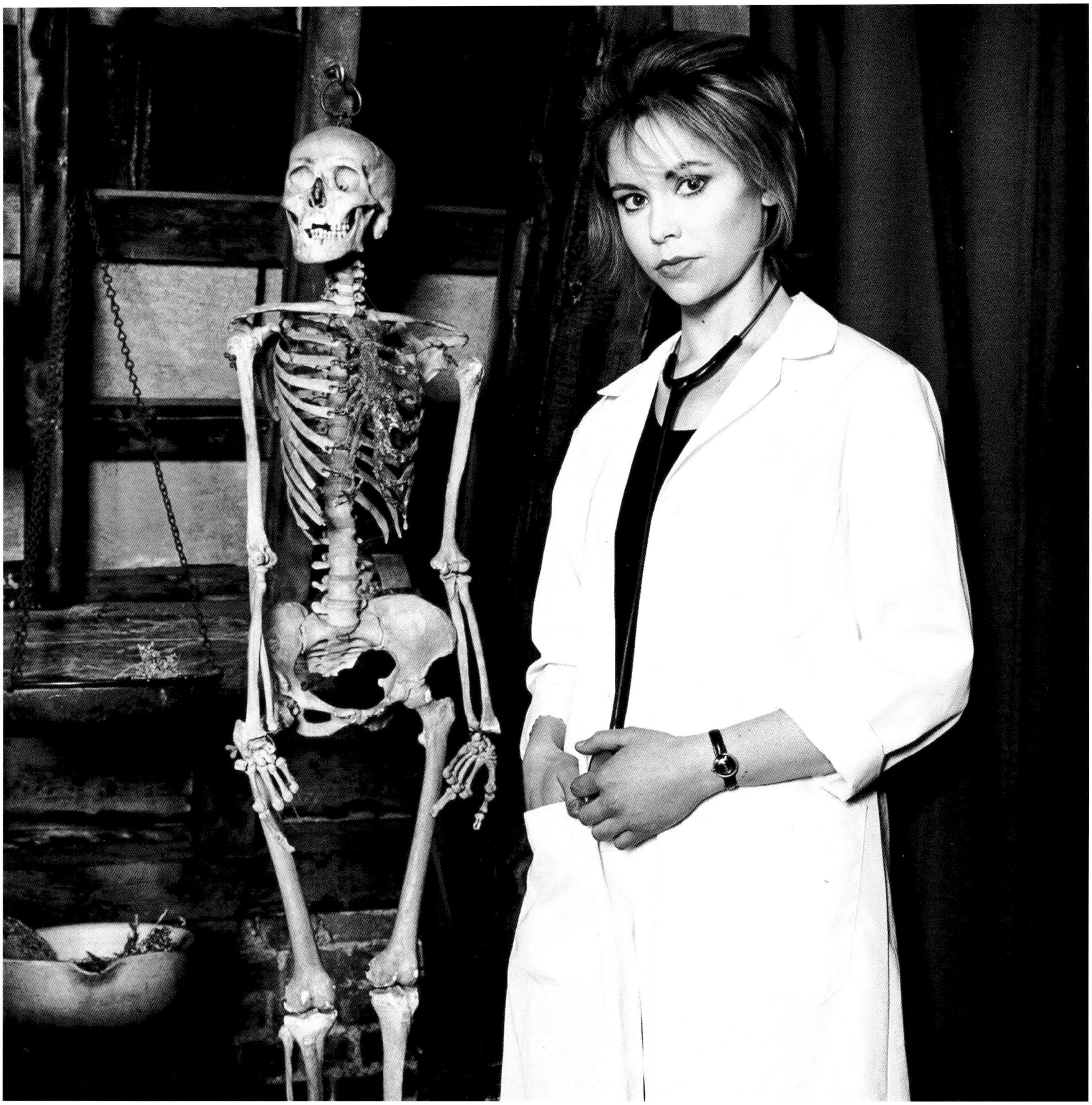

Dr Rachel Armstrong (photo: Nicholas Sinclair)

WHAT IS SCI-FI AESTHETICS?

DR RACHEL ARMSTRONG

The image of humanity is under threat. Having overthrown the notion of Divine Will as the force controlling our individual destiny, science sought to examine the natural world and to look at ways in which it could usurp the irrationality of 'natural selection' by taking control into its own hands. Using the techniques of experiment and objective rationalism, scientists provided the modern world with a plethora of devices by which humankind could overcome its inbuilt limitations – its faulty design, slowness to adapt, and ultimate mortality – without suffering guilt. Pure objectivity, without recourse to a moral judgement, allowed science to justify and practice previously unspeakable, blasphemous acts.

In *Frankenstein*, Mary Shelley warned of the dangers of these man-made interventions in the control of our destiny, using the imagery of electricity and dissection – fashionable scientific theories and techniques at the time. A century later Josef Mengele and others were recruiting human models from mental hospitals and concentration camps to test their theories of designer breeding, carrying out their abominations in the name of 'eugenics'.

But abominations and monsters are not only the result of 'unnatural' intervention: at war with themselves, our own bodies produce mutation and cancers, unaided by science. Each successful expression of a single protein is the triumph of one repressor or activator system over another. This battle begins even before conception when millions of unsuccessful spermatozoa die without ever reaching the mature ovum. Once the egg has been fertilised, there is programmed cell death in the embryonic tissue even prior to the birth of the child, and throughout our lifetime cells die, are replaced, or adapt to altered physiological demands.The final outcome of our own biological battle is death. What would it be like if we could keep our cells and tissues at peace, working harmoniously together?

The promise of leaving behind the restrictions of the old, natural biology and finding immortality in the ideal, self-healing body is gradually becoming more tangible. Scientists are advancing the understanding of the body in nanomolecular detail to create theoretical models of the ways our body mechanisms work and in doing so, proposing interventions to upgrade our biology. This collective aspiration towards physical perfection and eternal life is a culturally endorsed archetype akin to religious allegory – a quest for the Divine condition.

Reproductive sex and the laws of natural evolution as proposed by Darwin are simply not able to deliver this state of existence in our lifetime. However, sexual reproduction serves as the best existing means of extending our genetic life-span, which is the closest we can aspire to achieving an immortal status. We have socialised our expectations of sexual reproduction to such a degree that those who do not conform to these expectations (by engaging in fetishistic or nonreproductive forms of sex, or even by ageing) are considered alien.

As far as most of us are concerned, we are confined to the limitations of our biology and will have to make do with romanticised interpretations of our 70-year life cycle. During this period, most of us will seek immortal status by creating near-reproductions of ourselves through bearing children. Owing to the importance that we place on reproductive sex, we spend a great deal of our lives worrying about sexual function.

Sex plays so much part in our socialising and communication skills that our identities are shaped according to bipolar gender types. However, when we are no longer sexually reproductive or active we lose our primary motivation to interact with others and look forward only to senescence and death. This would not be such a drastic state of affairs if we could think about ourselves and our relationship to sexual activity differently. But we are DNA obsessed. Our whole philosophical approach to understanding identity is based on the scientific metaphor of genes. Darwin thought of these as the agents of inheritable characteristics, but although this metaphor serves as a useful tool for theorising about perceived evolutionary trends, it is not a practical mechanism for creating real change, nor is it totally understood.

The most realistic way of escaping our sexual function at the moment is through test-tube fertilisation techniques. Cloning is currently of great current interest, perhaps thanks to Dolly, the sheep who was created by Ian Wilmut and team at the Roslin Institute in Edinburgh by transplanting genetic material from a sheep into an egg from which the nucleus had been removed. The technique, which had never before been successful in mammals, used the genetic material of only one parent, rather than diluting the 'mother's' genes with a set of complementary male instructions. This does not bypass her mortality but Dolly is living proof that she can continue to produce identical sheep for as long as she has genetic material to harvest.

Cloning could be the best means of reaching immortality in this lifetime and would make gender identity simply another individual aesthetic attribute. If we are truly able to escape our sexual function, gender will serve a purely social function. The decision to be gendered or ungendered will not be a biological one but will reflect our desire for social and sexual interaction. Eroticising of other people, as individuals, rather than prospective biological investments, will sustain our motivation to engage in social activity during our extended lifetimes.

But despite the significant advances in the science of genetics including cloning, transgenics and gene harvesting, we are still no closer to being able to design a perfect body to our individual specifications by these means. We do not completely understand how genes are regulated, controlled or expressed. The study of developmental biology reveals that a whole variety of other cellular mechanisms comes into play in order to bring an organism into existence and these control rather than are controlled by the genetic code.

DNA is a stable, inert nuclear structure that does not actually design us but merely acts as a reference library that can be used by other molecules operating at a distance from it. Downstream of this a series of much more active and transient molecules such as the RNAs (Ribose Nucleic Acids), enzymes, hormones and proteins continually perform active work that gives form and function to the cell. If something goes wrong with these active molecules, the cell function is disrupted, despite a fully operational genetic code. The technology that promises to work at this level of cellular activity and cell-determination is the futuristic science of nanotechnology.

This new branch of techno-medicine promises autonomous machines or artificial proteins that will influence the operating instructions of the cell without interfering directly with the genetic code. Working by the laws of diffusion, these minuscule robots will potentially be able to police our bodies to ensure that all tissues are in perfect working order. They may even be programmed to improve our body design if we so desire. Some people have already had their bodies cryogenically preserved in the belief that when they wake from suspended animation the new nanotechnology will not only heal their old damaged biology but will give them a new aesthetic identity.

It is thus possible that we may be provided with the tools to direct a self-engineered evolution, but developments in nanotechnology still have a long way to go before they affect the population as a whole. More immediate benefit may be available to our ailing bodies from surgical intervention, the quickest and most direct method of manipulating the flesh. Although surgery has been traditionally used to rectify damaged body parts it is now increasingly being employed to augment our bodies in order to take control of our own identities and escape our biological determinism. In a society where image and appearance are the most important factors in the realisation of individual success, and as cosmetic surgery gradually becomes safer and cheaper through medical advances and new anaesthetic techniques, it is likely to have a dramatic effect on a large number of people in Western society. We will make more and more demands on our surgeon-sculptors and these demands may not simply be aesthetic ones, but social and political statements.

In an ideal world, perpetual servicing of the body would allow us to keep on living as each organ stops working. In practice, repeated surgery is not likely to deliver immortality by eternal replacement of worn-out body parts because there are problems with the compatibilities of the old and new structures. Silicone implants and metal prosthetics all have limited life spans in the natural body. The carbon-weighted immune system rejects 'foreign' bodies, resulting in a variety of downstream events that limit the number of interventions any one individual can undergo over a lifetime.

Descartes's metaphor of the body as a machine to house the human spirit has been taken literally by some philosophical activists. The first attempts to build mechanical life forms in the late 19th century gave rise to a number of strange mechanical devices such as a curious, egg-laying, mechanical duck. By the early 1930s, efficient mechanisation led to mass production and through assembly lines the human labour force politically became part of the machinery. In *Rosum's Universal Robots*, Karel Capek first coined the term 'robot' in reference to mechanical creatures that simulated the activities, if not the emotions, of human beings.

These artificial constructions have become part of modern mythology and have been brought to life as military, industrial and domestic servants. With the advent of intelligent and micro-miniaturised machines, the human-robot has become more physiologically detailed, mimicking its role model in behavioural terms far more accurately than in physical appearance.

The 'Tamagotchi' egg, an extremely popular Japanese toy, is an example of a successful robot life-form. Programmed to exhibit attributes characteristic of biological systems such as hunger and pain, this anthropomorphic creature allows human carers to interact with it through a rudimentary micro-miniature visual display unit and set of simple buttons.

Humanoid machines and their artificial biology have led to the speculation that the human body is merely a link in an evolutionary succession with mechanical life at the apex. This is perhaps wishful thinking, since, according to Hans Moravec, Professor of Robotics at Carnegie Mellon University, robots can adapt far more easily to hostile environments, such as space and deep seas than fragile human bodies. These habitats may be the only ones available in the future, to support social populations in the advent of global terrestrial overcrowding.

But these eternal riddles of life, death and development are not simply the province of science.

A number of visionary artists are revealing to their audiences marvellous but credible versions of another material reality, utilising a sci-fi narrative and a scientific methodology to record and rationalise their ideas. This framework allows them to operate within the limits of scientific possibility but does not force them to surrender their artistic ambition and imaginative exuberance to its conventions. Invading the laboratories, stealing scientific techniques in order to expose our complacent dependency on scientific progress, they are challenging the truths presented by the scientific establishment.

For these artists the human body is not merely an aesthetic object but also acts as a locus of political power and wealth. Collectively they imply that the human body as a divine subject or a pure product of 'Nature' is false or imperfect, expressing a desire to destroy our current biological design and replace it with a more advanced form.

Some, like David Bowie, Simon Costin, Peter Gabriel and Mariko Mori effect their metamorphosis using a range of different media such as video, photography and computer imaging technologies. Predrag Padjic achieves transformation by making skin-like costumes that are an extension to the body, as does Kenjie Yanobe who protects us from our hostile, polluted planet with his space suits and body wear. Tina Keane suggests an opposing scenario: that the human body will not remain distinct but will become absorbed by the environment, a mere element of a future landscape. While Jake and Dinos Chapman and Matthew Barney conjure up mutant sexualities and genders, others such as Orlan, Stelarc, Franko-B and Ron Athey literally transform their own bodies.

Self-destruction, mutilation and dismemberment of the human body in the form of surgery and hybridisation with machines are being used by many contemporary artists as distressingly real techniques to mould the flesh into new expressions of humanity. Unlike the Viennese Actionists who inflicted their own physical pain on the audience through violently sadomasochistic imagery in an attempt to emphasise their humanity, the current body artists are manipulating their flesh in conjunction with radically advanced medical appliances, to present a non-suffering, super- or post-human body.

These artists witness their own mutilation from a position that seems to be outside of their own body. In their work, the body becomes a cadaver unperturbed by the direct physical insults inflicted upon it by the will of the artist/creator. But is playing the role of both Dr Frankenstein and his monster a valid method of artistic expression, or is it an antisocial, self-indulgent, sadomasochistic act?

By transforming their own external images the sci-fi artists are creating a series of self-portraits where their own flesh, sinews and organs become the canvas. Their new humanoid life forms are portrayed as possible successors to the present human species. Each suggests a potential future for humanity, but whatever their political or artistic motives, they use their selected techniques only as a tool for self-reinvention. This experimental process, may be seen as a 'pure' artistic act, or as a crime. But these abominations are actually no more disfiguring than those murderous strategies employed by nature itself (investigated here by Roberta Graham and Kathy Acker). Natural selection and genetic determination ruthlessly deal some people 'fairer' combinations of functional and physical attributes than others. The metamorphoses and variations displayed by these artists depict extreme possibilities that mimic an evolutionary progression.

Cindy Jackson, who has achieved a remarkable transformation by undergoing a massive series of operations to become more beautiful, demonstrates that physical beauty equals power and wealth. By contrast, Orlan employs cosmetic surgery to challenge the status quo. Her disturbing, confrontational 'Carnal Art' uses living dissection as a means of provoking questions about the tyranny of our anatomy and the aesthetics of the body. The most powerful message she conveys through these drastic, radical acts, is that an individual can gain control over their face and body to reflect their true nature. Elizabeth DYN is creating her vision of 'The Face' to achieve harmony between body and soul. When it is completed she will embark on a spiritual journey, performing good deeds and preparing a new culture for the millennium. Working with an anti-biological and pro-mechanistic vision, Stelarc is literally becoming a cyborg – gradually replacing his body parts with mechanical devices – and is currently exteriorising his nervous system to link with the Internet.

Despite the sci-fi artists' attempts to depict plurality and encourage self-expression in new humanoid life forms, a consensus future human blueprint may actually be emerging. In his article 'The Alien Image' Michael Buhler warns of this stereotype infiltrating our collective consciousness. This large-headed, almond-eyed creature, technologically sophisticated, genderless, objectively rational, psychic, multi-lingual and immortal, may be our successor.

Evolutionary forces are at work: natural laws, in combination with man-made technologies and unknowable or alien forces are compelling us to change. We may wish to deny or resist it, but it is happening to us now. The articles and images collected in this issue of *Art & Design* will reveal progressively stranger configurations of the body until a point is reached where the human element is no longer evident.

The image evoked by accounts of alien abduction, where alien doctors and scientists conduct clinical experiments on dazed and confused abductees, may be seen as a metaphor, formed in the collective consciousness, for the self-assessment that humanity is carrying out on itself, uncertain of its own future. The sci-fi artists do not placate or comfort us in our lack of self-recognition as we diversify. Their images help us to accommodate or adjust to new and existing transformations. Through their portraits they offer ways of bridging the alienation imposed by dominant conventions of human form. Proactive rather than retrospective, sci-fi artists suggest that we are able to play a part in shaping our species in preparation for the time when we can look at ourselves from the outside and objectively assess our humanity.

SIMON COSTIN

Burning

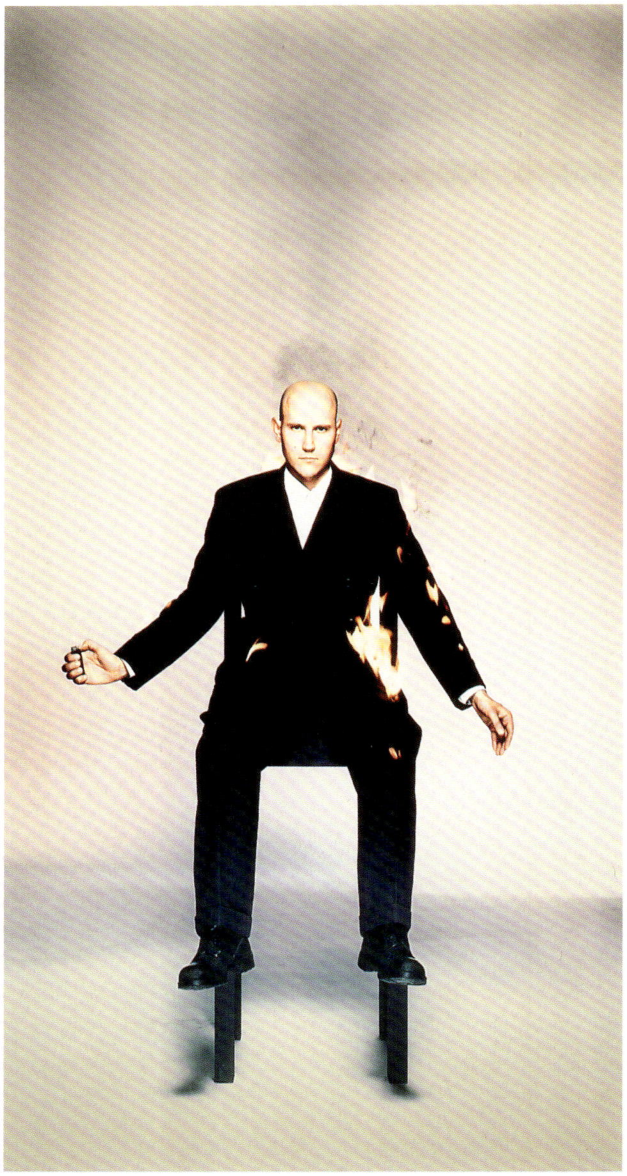

Evolution has not been a gradual process: the development of the eye, the extinction of the dinosaurs, the arrival of mammals, the attainment of dexterity in humans occurred as a series of revolutionary step-like jumps in reaction to the changing environment.

Many of the cataclysmic events of history have been the result of the will to achieve radical transformation – to start life again through auto-destruction. A new millennium precipitates anxiety about the extinction of the human race. Doomsday cults with their mass suicides – such as the group, Heaven's Gate, who thought that comet Hale-Bopp was a celestial sign, from the 'Evolutionary Level Above Human' – have featured prominently in the media and are likely to increase in the next few years: there is a popular public desire to self-destruct and start again.

For his series of photographs, *Burning*, Simon Costin set himself on fire using special stunt techniques. The portraits were then given an even more disturbing, hyper-real quality by computer enhancement. Throughout the gruesome process he stares back at the viewer dispassionately, refusing to provide clues to the cause of combustion. Nor does he satisfy his voyeurs with proof of his death, but holds us in spellbound suspense.

It is difficult to tear oneself away, yet almost unbearable to behold this artistic martyrdom. In many ways it is reminiscent of the experience of watching television coverage of the Bradford football stadium disaster. The image of one particular man, featured with each news item, became an icon of the tragedy. The video footage showed him failing to climb over the restricting wire fence, burning as he fell. There was no possibility of intervention: we watched him burn.

Costin's triptych encapsulates the ritualism and repetition of the image of that immortalised victim whose burning body symbolised the greater collective pain suffered by the victims and their families. A magical quality is conferred by the artist's consumption in flames: a kind of transcendence of the flesh more significant than the actual horror of self-destruction. The reconstruction of this archetypal myth as an

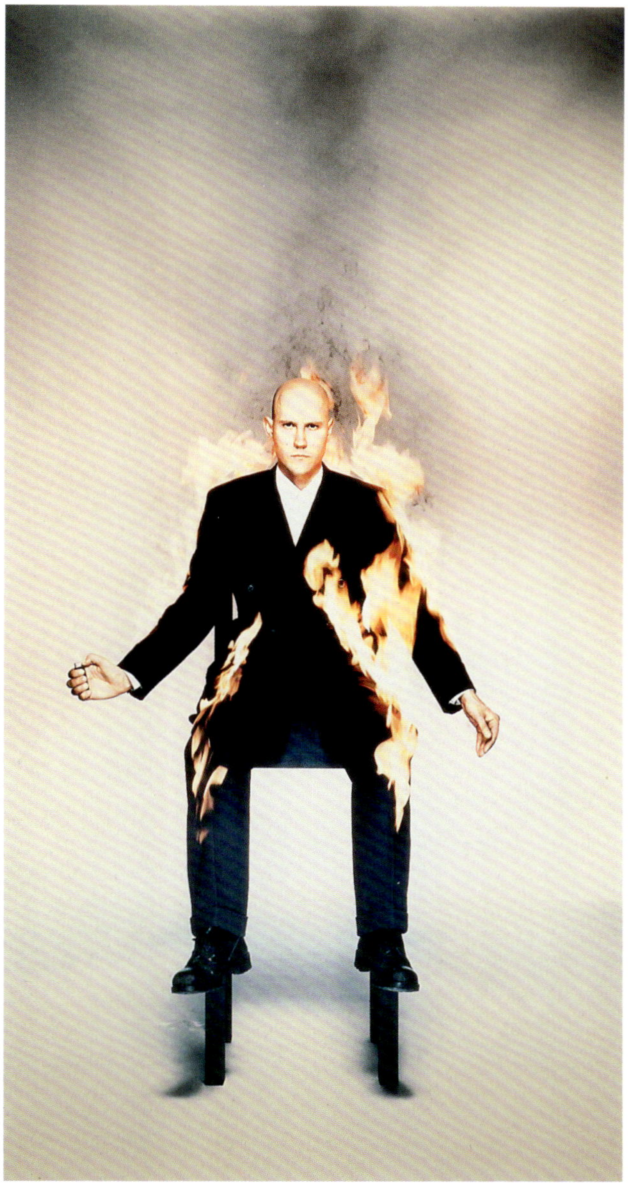
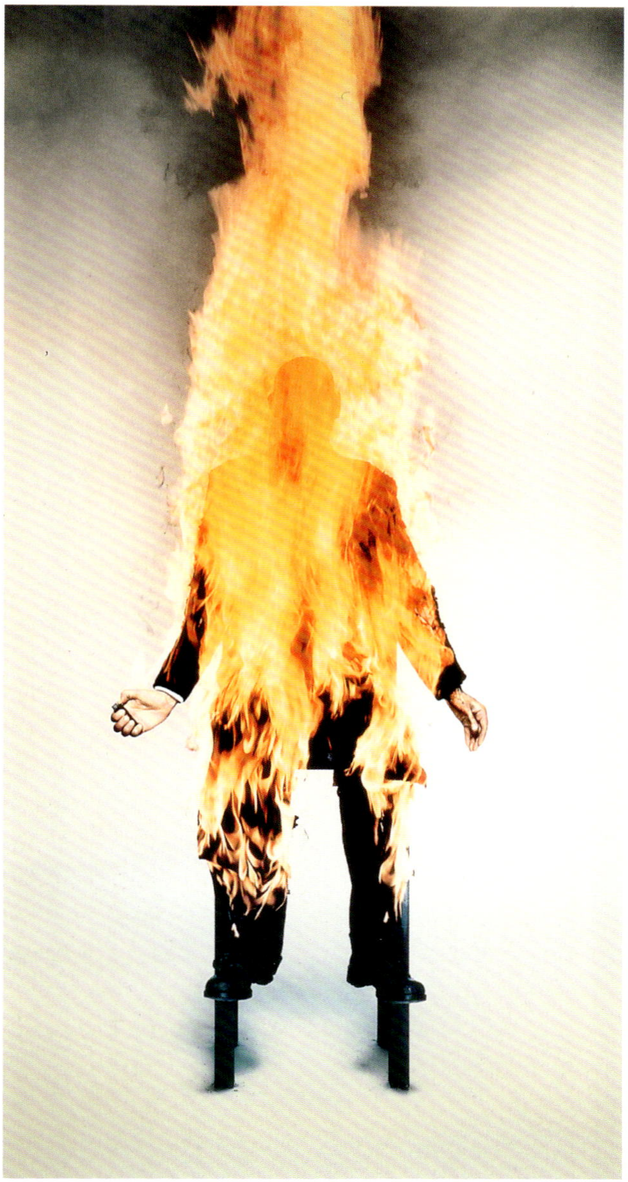

ordeal and a fantasy symbolises to the gallery audience the possible victory of the mind and spirit over the laws of nature.

This imagery of radical transformation by fire has often been used in art, most notably in science-fiction films. Sometimes fire may bestow superhuman properties, such as immortality (as in the cult classic *She*, where the goddess remains beautiful and vital by bathing daily in the eternal flame). At other times it is used in order to effect the radical destruction of a culture, such as in Ray Bradbury's *Fahrenheit 451* where the fireman hero gradually discovers his passion for literature in a world where books are illegal. Each time a book is consigned to the flames it is as if an act of murder has been committed.

Costin exploits the paradox at the heart of this symbolism by maintaining his physical composure in the midst of the conflagration, the beauty of his dissolving flesh glamorising the distressing experience and anaesthetising our horror. He offers us images associated with martyrdom, a state that confers benefits to the victim in the afterlife, and his ordeal by fire promises the reward of transformation into the superhuman.

Our exposure to the techniques of digital image manipulation has encouraged us to accept this fantasy as part of everyday life: we expect horror to remain fantastical. New technologies such as Virtual Reality and console games glamorise death, promising immortality and the possibility of transformation: we are able to die many times and come back as new characters or inhabit different bodies. The virtual characters with whom we identify mesmerise us into forgetting our role in consensual acts of violence and death. Costin, however, found his inspiration for these transforming horrors in the real world, when, in 1993 he witnessed a Bosnian protester setting himself on fire outside the Houses of Parliament.

In *Burning,* Costin sets the scene for our cultural auto-destruction, suggesting that it may be new technology together with our spiritual, political and creative complexities that will bring about the anticipated millennial evolution. *(RA)*

Burning *(triptych), 1996, Cibachrome prints, each 240 x 120cm (photos: Nick Knight)*

ANDRES SERRANO

The Sea of Possibility

JIM HAROLD

> *What I did with the cadaver was to cover his eyes, his ears, his nose, his mouth. I slapped his face, I blew the hair on the back of his neck, I pinched his arm . . . I was attempting to get to that person, not that individual, but that symbolic person represented by the cadaver, who is in me.*[1]

This quote is taken from an interview with the artist Matt Mullican. His words are both a pragmatic description of his interventions into the stillness of the corpse, and a disclosure of another urge – his attempt to make contact with a symbolic figure, held in common by the viewer and the object, representing the sense of his own nonexistence. This is a difficult zone to enter, lying at the limits of rational understanding; an area heavily overlaid by the complex emotions of absence and loss. Mullican is not concerned with the corpse as an object that triggers pain or grief – though such thoughts must hover in the background – but as a threshold to the 'unknowable'.

The dead body arouses our fear, forcing us to translate its presence into something we can understand: a system of codes and signs that situate it within culture. Western societies live in uneasy relation to the world. Whilst emphasis is placed on the material, the possibility that a metaphysical dimension may exist has never been fully erased. This is clearly illustrated by the complex world of 17th-century medicine, as revealed in Rembrandt's *The Anatomy Lesson of Dr Nicholaes Tulp,* 1632. In this painting the cadaver has been partially and theatrically opened, not just in the name of science, but also in the pursuit of spiritual and social morality. The identity of the corpse as a petty criminal gives a clue to this. Prevalent at the time was the notion that recidivists could be purged of their sins and turned towards an ultimate good, if not in life, then certainly in death, under the anatomist's hands.

Everything in the painting points to the relation between the anatomist's hand and the cadaver's own arm and hand. Dissections normally start with the removal of internal organs, and end with the extremities. In a real dissection the body would hardly be recognisable as human by the time the hand was reached. Here, not only does the operation start there, but the corpse's left arm appears to have the musculature of a right arm. The explanation would seem to be that:

> Tulp and his surrounding surgeons are investigating the human hand, the organ which (along with the intellect) differentiates the human as a tool-making animal . . . So, Tulp mediates with his forceps between the written authority of the text (shown in the lower right foreground of the painting) and the corporeal reality of the human body (the corpse is a corpse and nothing more), and the immortality of the soul (the body is a 'temple' created in the image of God) endowed with the cognitive faculty, as the unique structure of the hand indicates.[2]

The hope that the secrets of God's world, the mystery of life, would be fully understood has not proved to be sustainable in any real sense. The impossibility of intellectually grasping death, the 'unknowable', has given rise to a system of elaborate avoidance techniques. In his book on mortality and immortality in modern European cultures, Zygmunt Bauman argues that while humans cannot undo either the knowledge or the inevitability of their mortality, the mechanisms of contemporary culture territorialise death and try to offer the palliative of a secularised quasi-immortality – whether through the mechanisms of the life-support system, cryogenics, or the cult of the celebrity.[3]

Looking at death within modern European cultures, the journalist Michael Ignatieff comments:

> Cultures that live by the values of self-realisation and self-mastery are not . . . especially good at dying, at submitting to those experiences where freedom ends and biological fate begins. Why should they be? Their strong side is Promethean ambition: the defiance and transcendence of fate, material, and social limit.[4]

Part of this drive to immortality leads culture to minimise the presence of death or inversely, to valorise it through a variety of forms of institutional idealisation ranging from the aesthetic to the medical and legal. In this regard, hospitals (and doctors – whose role was once to provide the dying with a good death) attempt to negate death as part of a strategy to assert the life-extending potential of their technocratic systems. Mortality in such a system can only be an enemy, an obstacle to be overcome in a futile all-out war.

It is against this backdrop that Andres Serrano's *The Morgue (Cause of Death)* series should be viewed. These photographs exhibit the failure of the system to transcend the realities of death, and exploit our human, deep-seated disquiet at the sight of it. The abrupt demise of these nameless subjects – whether through murder, drowning, incurable illness, or suicide – underlines the brutality of death. The final assault on our senses is the utter uselessness of their corpses.

Despite the apparent objectivity of the photographs, Serrano attempts to ease our fear by making reference to familiar art historical motifs and to the many symbols of transcendence and redemption in Christian, and in particular, Catholic iconography. *Infectious Pneumonia,* 1992, is a tightly cropped photograph of a woman's head, half covered by a rich red cloth and shown in profile. The single strand of hair that has caught in her closed lips carries such pure poignancy that the viewer blurs the boundary between death and sleep. *Rat Poison Suicide II,* 1992, is a photograph of a foot that exhibits a livid gash reminiscent of the marks of crucifixion.

But whilst many of the works inhabit the parallel, if uneasy, zones of material death and its cultural depiction as repose or release, a much more difficult strand of meaning is woven into some of the other images in this series. Works like, *Hacked to Death II,* and *Death by Drowning II* increase the viewer's discomfort by negating the easy escape routes associated with the idealisation of human transcendence. The undercurrent revealed in these unrelenting images of dead flesh is that the moment of death can neither be easily sidestepped nor glossed over. They are a blunt depiction of mortality.

Serrano's photos do not penetrate the body for meaning in the way the anatomist does, rather, they inhabit its surface, capturing the point at which the dead body becomes an image of its former living self:

> Man is made in his image: this is what the strangeness of the cadaver's resemblance teaches us. But this formula must first be understood as follows: man is unmade according to his image. The image has nothing to do with signification or meaningfulness as they are implied by the world's existence, by effort that aims at truth, by law and the light of day.[5]

This is, I suspect, where Mullican hopes to find his symbolic figure: the point where the artist and the viewer acknowledge, if only for a moment, their own nonexistent being. These two works certainly do not create a space for idealised revelations, or for possible transcendence. They depict a shocking moment between, a moment described by Julia Kristeva in her complex meditations into the saturnine nature of Hans Holbein's *The Body of the Dead Christ in the Tomb,* 1521:

> Cut off from us by its base but without any prospect towards heaven, for the ceiling of the recess comes down low, Holbein's Dead Christ is inaccessible, distant, but without a beyond. It is a way of looking at mankind from afar, even in death – just as Erasmus saw folly from a distance. It is a vision that opens out not on glory but on endurance. Another, a new morality resides in this painting.[6]

In this work of the Reformation period, Holbein addresses the European dilemma of religious representation by voiding the work of any possible transcendent light: this is a Christ who is simply not going to rise from his tomb. It is no coincidence that *Dead Christ* appeared at the very moment when cultural conditions were ripe for the concept of the individual to develop. However, it is important that the painting – the first to give a visual form to the autonomy of the individual – is also at pains to underline the resultant isolation of that individual from the comforting contact with the Divine.

While Holbein's *Dead Christ* seems to illustrate both the imminent and long-term consequences facing the new concept of the individual as a being who has become both separate from the Divine and existential, Serrano presents us with the further vision of the total isolation of that idealised individual. It is a vision of an individual who has been offered the autonomy to control their destiny (even by suicide) so as to be able to make contact with Divinity, but who is ultimately denied the necessary belief systems to achieve it.

The art work is thus denied its role as a conduit for the illusion of transcendence and immortality. This is not as negative as it may appear, for whilst in these works Holbein and Serrano actively negate the concept of the ideal, they do stress a new relationship with the moment of severance itself, posing questions about the nature of meaning, representation and of our own existence and nonexistence. By removing the palliatives normally sought as a result of human desire (whether that desire is driven by the quest for knowledge or by belief), they project the viewer into the spaces of the image that open out between reality and ideal:

> '. . . and he pressed the blade to his smooth throat and let it dip in . . . dip in to the sea of possibility. . .'[7]

It is as if, through the image, the knife is being pressed to our own neck as we stand on the edge at the moment of severance, watching the corpse that we will become in our turn – whether we are the anatomist, the patient, the victim, the murderer, the suicide, the artist or the viewer.

Notes

1 Interview with Matt Mullican by Michael Tarantino, *Parachute* (Canada), Oct/Nov/Dec, 1992.
2 Jonathan Sawday, *The Body Emblazoned,* Routledge, London, 1995.
3 Zygmunt Bauman, *Mortality, Immortality, and Other Life Strategies,* Polity Press, Cambridge, 1992.
4 Michael Ignatieff, 'Modern Dying', *The New Republic*, Dec 2, 1988.
5 Maurice Blanchot, 'The Two Versions of the Imaginary', from *The Space of Literature*, 1955, Ann Smock, (trans) Nebraska University Press, 1982.
6 Julia Kristeva, the essay 'Holbein's Dead Christ' from *Black Sun, Depression and Melancholia*, Columbia University Press, 1989.
7 Patti Smith, 'Land' from the album *Horses*, Linda Music Corp, 1975.

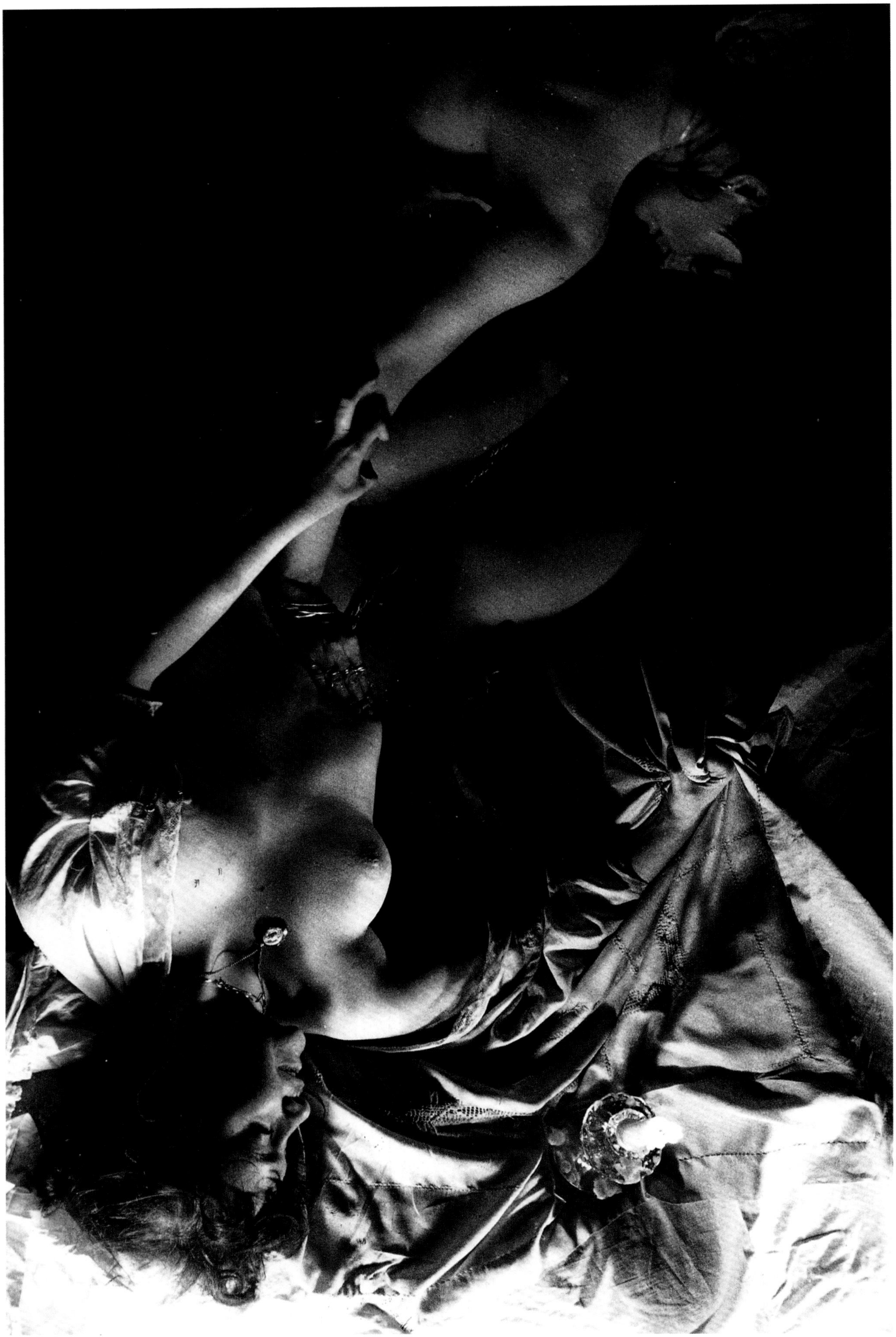

ROBERTA GRAHAM

Fallen Angel

A flash of lightning illuminated the object and discovered its shape plainly to me; its gigantic stature and deformity of its aspect, more hideous than belongs to humanity, instantly informed me that it was the wretch, the filthy daemon to whom I had given life . . .

'Remember, that I am thy creature; I ought to be thy Adam; but I am rather the fallen angel, whom thou drivest from joy for no misdeed.'

Mary Shelley, *Frankenstein*

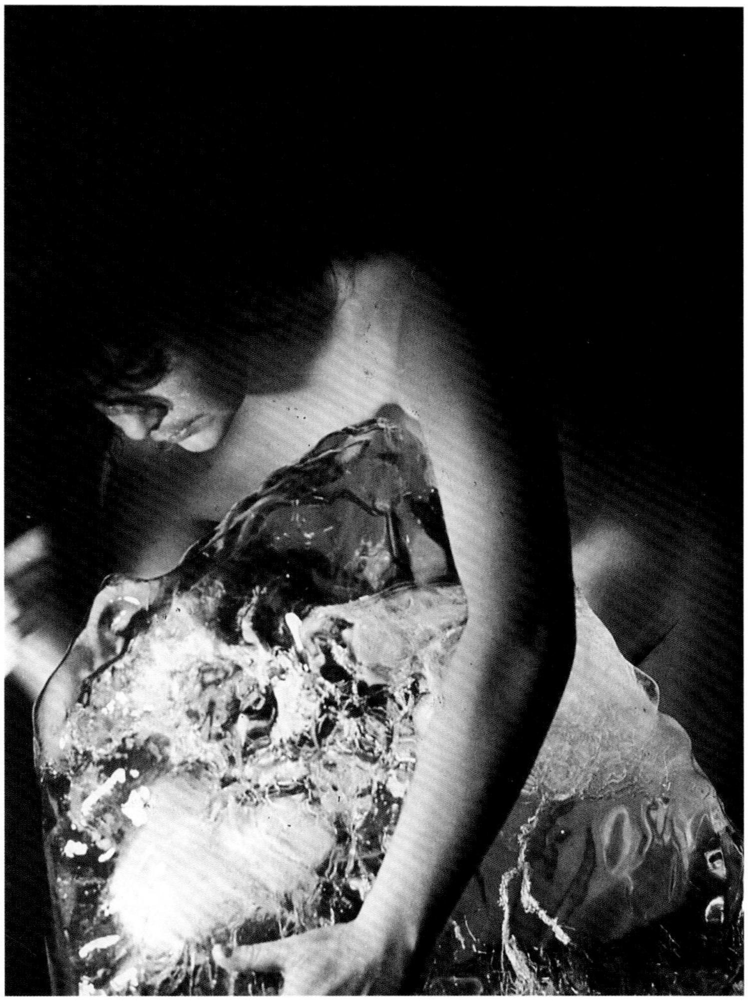

Fallen Angel, *hand-painted photographic panels, 1.3 x 1.22m; ABOVE: Panel 1; OPPOSITE: Panel 4*

Mary Shelley's novel, *Frankenstein,* begun in 1816 at the Villa Dioadati, near Geneva, expressed the complex mythology of creation and destruction. The corpse-creature, brought to life by an undeveloped modern science, is born with an innocently murderous nature, the monstrous miracle of its creator.

At the heart of Mary Shelley's story lies a fearful entwining of birth and death that was painfully real to the women of the period. Death in childbirth was a common occurrence, often the result of natural complications combined with crude or ineffectual medical techniques. Even if the woman survived childbirth, her baby's life expectancy was pitifully short.

Shelley's birth resulted in the death of her mother, Mary Wollstonecraft, from an infected womb, and her own attempts to have children were blighted by tragedy. Her first child died before it even bore a name, returning to her in recurring nightmares in which she felt its cold body revive in her arms. Two more children failed to survive early infancy and finally, Shelley haemorrhaged badly during a miscarriage.

Fallen Angel is a series of eight, hand-painted photographic panels and a 20 minute tape/slide work, exploring the anguish of women in an age where birth was closely shadowed by death and where the outcome of creation could therefore be an act as unjust as destruction. The monster – the Fallen Angel – is an embodiment of this anguish, symbolising the terror and revolt of the creator/mother. The imagery draws parallels between a nascent modern science, which sought to regulate and control matter, and the symbolism of alchemy, which attempted its transmutation. *(Roberta Graham)*

THE GIFT OF DISEASE

KATHY ACKER

I am going to tell this story as I know it. Even now, it is strange to me. I have no idea why I am telling it. I have never been sentimental. Perhaps just to say that it happened.

Unlike most medical stories, all the horror in this one occurs at the beginning. Gradually there will be an end to the fear.

In April 1996, I was diagnosed with breast cancer. I had had a history of breast lumps; until this time, none of them had been malignant.

Most of what happened occurred in the United States. I don't know whether or not conventional cancer treatment is handled differently elsewhere.

A biopsy revealed that the mass involved was less than 5cm in diameter. Since cancerous cells were found at the rim of the extracted tissue, my surgeon told me that I had two choices: either a lumpectomy with radiation or a semi-radical mastectomy without radiation. Given the size of the mass and the findings of his examination, he said that, according to statistics, there was only a 30 per cent chance that the cancer had metastasised. If it hadn't, I wouldn't need chemotherapy.

I was terrified of cancer. I feared chemotherapy more.

At that time, I was working as an adjunct professor at an art college and so did not qualify for medical benefits. Since I didn't have medical insurance, I would have to pay for everything out of pocket. Radiation on its own costs $20,000; a single mastectomy costs approximately $4,000. Of course, there would be extra expenses. I chose a double mastectomy, for I did not want to have only one breast. The price was $7,000. I could afford to pay for that. Breast reconstruction, in which I had no interest, begins at $20,000. Chemotherapy, also begins at $20,000.

According to statistics, in the Bay area of California, where I was living at that time, one out of seven women was being diagnosed as positive for breast cancer. A friend who is a noted nutritionist privately told me that experts expect these statistics to rise to one out of three and that The Center for Disease Control in Atlanta, Georgia, has been called in to investigate. None of this has as yet been let out to the media.

Even according to officially approved statements, the rates for breast cancer in northern California are the highest in the world. Moreover, it is not known why so many women, the majority of whom are white career women, are developing breast disease.

Breast cancer, in the realm of Western medicine, is big business. The two largest industries in the United States are the weapons industry and the medical industry; cancer research and care are a mainstay of the medical business.

Having been diagnosed, I consulted both my acupuncturist, who was also an MD, and my nutritionist. I had been working with both men for about six years. The latter put me on a high anti-oxidant diet. The former told me that there was nothing he could do: that acupuncture does not deal with cancer.

The double mastectomy itself took place three weeks later, at the very end of April. It was the same hospital where the biopsy had taken place, one of the finest in San Francisco. Even the waiting-room into which I was ushered, after I had prepaid all my costs, felt warm and safe. I spied conveniences: a phone and a TV, a closet in which to hang my street clothes, a private bathroom. The nurses who visited me in order to administer the necessary preliminary tests and preparations were both kind and friendly.

One of these nurses told me that my operation had been moved back an hour. Ten minutes later she wheeled me, in a wheelchair I didn't need – for I felt healthy to myself – through halls, down an elevator, through more halls to the edge of a second elevator door. The walls and decoration around me were lovely: wood and dark red colours.

Now a nurse whom I hadn't seen before brought me through two huge doors adjacent to the elevator into an unheated hall. Various green-capped and green-clothed figures moved by me.

While I was here, one of the green figures introduced a pre-anesthetic into the IV fluid that was dripping into my veins. As soon as she inserted the liquid, I felt cold creeping around the base of my skull, eating at me. My brains were nauseous. I knew that I didn't want to be here. Then I knew that I couldn't escape because my mind had been changed.

While this was taking place, a green, shower-cap-like thing similar to those worn by the figures was put on my head. It fell over my eyes: I could no longer see. Scotch tape was wound round what jewelry couldn't be removed. I was being reduced to something I couldn't recognise.

The next room was huge and colder than the hall. In its middle there was something that was partly a table and partly a bed. Machines that looked like Dr Seuss animals were connected to it. I was told to climb up onto it and lay on my back.

I want to describe as exactly as possible what it is like to experience conventional cancer medicine. However, I am omitting the more horrific details. Thick straps were placed around my arms and legs, then buckled tightly.

I remember asking, 'Why are you doing this?'

'Because we don't want you to harm yourself.'

While I was trying to figure out how I could hurt myself if I was under sedation, red suction cups connected to steel cords were placed on my torso. One of the figures around me asked me when I had last eaten. I replied, before midnight as I had been instructed. They asked about drinking. I remembered I had had a sip of water when I had woken up. The person who was asking the questions and taking my replies down in a notebook told me that I was in danger because, having sipped water, I might vomit during the operation, due to the anesthesia, and choke on my vomit.

I was scared.

The first thing I did when I came back to consciousness, I remember, was try to stand up because I wanted to get out of this hospital as soon as possible. I wasn't staying the night, even if I wanted to, because only people with medical insurance were allowed to do that. But I couldn't stand up. Downstairs my lover was waiting for me in a taxi; an orderly wheeled me to the vehicle while I fought against my nausea.

My surgeon had told me, prior to this operation, that I could begin exercising the day after it occurred. But two days later, my left arm still wouldn't move. On that day I had an appointment to receive my pathology report at the surgeon's office. I drove there on my motorcycle, with my lover on the back.

In the medical office I was told that my lymph nodes were registering cancer. I remember the conversation as if it were a play:

My Surgeon: I want to explain this situation to you. All of us are going to die . . .

Me (interrupting him): You said that there was little chance that cancer had spread to the lymph . . .

My Surgeon: That was before this operation. Now things are different. I will explain so that you can understand. All of us are going to die. Some of us are going to die in 20 years; some in five . . .

Me (interrupting him): Are you telling me I'm about to die?

My Surgeon: No. I'm not telling you you're about to die. I want you to understand this situation. There's a good chance you'll walk out of here and never have anything to do with cancer again.

Me: What sort of chance?

My Surgeon: According to statistics, at your stage of cancer, there's a 60 per cent chance of non-recurrence. If you do chemotherapy, according to the most reliable statistics we've got, those chances go up to 70 per cent.

My Lover: Doing chemotherapy raises her chances only 10 per cent?

My Surgeon (sadly): We don't know any other way to deal with cancer at her stage except chemotherapy.

Me: Tell me about cancer and lymph nodes. If my cancer's so advanced, how come there wasn't any other cancer in the breast tissue? Aren't the lymph nodes the body's filter? Couldn't it be possible, since I've been on a super-high, anti-oxidant diet ever since the lump was discovered, that the lymph nodes have been doing just what they're supposed to do? Couldn't they be registering cancer because they've been cleaning out the diseased cells?

My Surgeon: Unfortunately, studies – and I've read the best – all indicate that there's no connection between diet and cancer. The same is true with regard to environmental pollution. The truth is that we don't know what causes cancer.

I have recounted this conversation as accurately as I can in order to make one point. The point is a realisation. As I walked out of his office, I realised that if I remained in the hands of conventional medicine, I would soon be dead, rather than diseased, meat. For conventional medicine was reducing me, quickly, to a body that was only material, to a body without hope and so, without will; to a puppet who, separated by fear from her imagination and vision, would do whatever she was told.

I'll say this in another way. When I walked out of that office, I thought that I might be about to die, to die without any idea why. My death, and therefore my life, would be meaningless. The reduction of all that one is to materiality is a necessary part of the practice of conventional Western medicine.

Actually what I became was this one thought: 'I want to live'. To live was to stay alive and not be reduced to materiality. There was no way I was going to go through chemotherapy.

I never got in touch with that doctor again except to pay his bill.

My search for a way to defeat cancer now became a search for a life and death that were meaningful. Not for the life presented by conventional medicine, a life in which one's meaning or self was totally dependent upon the words and actions of another person, even of a doctor.

I had already learned one thing, though I didn't know it at the time. That I live because I believe; that belief is equal to the body. The hardest part of my cancer was walking away from that surgeon and from conventional medicine. Belief in conventional medicine, in what our doctors tell us, is so deeply engrained in our society that to walk away from conventional medicine is to walk away from normal society. Many friends phoned, crying and yelling at me for not doing chemotherapy.

My friends' confusion mirrored that of the medical community. Today there is a health crisis in the United States and it is twofold. While epidemics such as cancer, AIDS, and immune diseases in general – whose causes, according to the practitioners of conventional medicine – are unknown, sweep through the populace, the industry is less able to care for those who are inflicted with them. This is not only because there are no known conventional cures, but also because medical insurance, as it has been set up, covers too few people, because there are not enough doctors, nurses and medical facilities for all but the rich. Faced with the enormity of what is happening, many doctors, such as my surgeon, say, 'We might as well do chemo because we don't know what to do.' The doctors will not admit that they are profoundly confused.

Even though I walked away from that surgeon, I had no idea where to go.

I turned to the man who had been my nutritionist for five years and asked him for the name of the best nutritionist in the city. The latter read my pathology report and said, 'There's only one way you can beat your cancer.'

'What's that?'

'You have to find out what caused it.'

He also advised me to pray.

When I walked out of that office, I was mad. Nobody knew what caused cancer. The incident remained in my mind.

I was looking for someone to whom I could turn, rather than looking inside myself.

Impelled by the fear that my time was limited, I began to learn as much as I could about alternative or adjunctive cancer therapy. The doctor who was my acupuncturist at that time, when I pressed him, informed me that the most accepted alternative cancer treatment was the Gerson method. Until the advent of the macrobiotic diet, the Gerson had been the best-known nutritional therapy for cancer in the United States.

In November, 1946, the American Medical Association openly attacked Dr Max B Gerson: they subsequently destroyed his professional reputation and denied him malpractice insurance. Gerson was forced to move to Mexico; today the Gerson Institute, directed by Charlotte Gerson, his daughter, operates partly in Bonita, California, and partly in Tijuana, Mexico. Most researchers and practitioners of alternative cancer therapies who have become nationally prominent have in one way or another been harrassed or directly threatened by the American Medical Association. Dr Stanislaw Burzynski, one of the most prominent cancer researchers in the United States today, began his investigations by studying differential patterns of peptides. After a congressional hearing at which his cancer recoveries were called 'miracles', he found himself outside the medical establishment and therefore without funding possibilities. In the United States, major criminals usually undergo two grand jury hearings. Dr Burzynski has endured five such investigations of work that many in the field, including Ralph W Moss, author of *The Cancer Industry*, consider brilliant.

Cancer is business.

If it is difficult to gather information about alternative cancer therapies, it is more difficult, in fact, impossible, to evaluate each therapy. To walk away from conventional cancer treatment is to find oneself in a no-man's land between legality and illegality. After two weeks of intense research, I was more desperate than when I had started.

The nutritionist's words, 'You must find out how you got cancer' were a thorn in my brains.

At this point I walked completely away from the society I had known. I took a leap of faith, turned to Frank Molinaro, a psychic with whom I had worked for about a year, and asked him for help. He referred me to Georgina Ritchie.

As I turned to Frank and then to Georgina, I saw myself – a person faced with her own death, and the meaninglessness of that death – turning to psychics, and I laughed.

Though I had known Georgina Ritchie for a year before I consulted her about my cancer, I hadn't known much about her. She had followed in the footsteps of her family, as was expected of people of her class. Graduating in economics from Barnard College, then from the sister school of Columbia University, she had married 'a proper husband' – a well-known Hollywood film director.

'I had been told, as all good girls did, to share in my husband's work. So I began to learn photography.'

When a professor who worked at the Humanist Psychology Institute in San Francisco invited her along on an expedition to what used to be the Inca empire, then elsewhere, to meet psychic healers, the Smithsonian Institute assigned her to take pictures of all she was going to see.

'In Mexico City, I saw a kidney transplant done with a rusty knife. The knife went through the spine, severing it. I saw the bad kidney taken out and the good one inserted. After the operation, since it was a psychic one, the patient just walked away. She hadn't felt a thing. As far as I know, she's still alive. You know, I have copies of all those pictures. I watched a healer who has clinics in Sao Paulo and elsewhere perform a psychic brain surgery. He used herbal concoctions; I saw him close the wounds with his own hands. Then, there was a healer in the Brazilian jungle who incorporated the multiple egos of schizophrenics.'

'Have you always been interested in shamanism?'

'Before I went on this trip, if anyone had told me I would become concerned with all this stuff, I would have been horrified! The only thing that prepared me for what I now do was a healthy distrust in Western medicine. My dad was a doctor.'

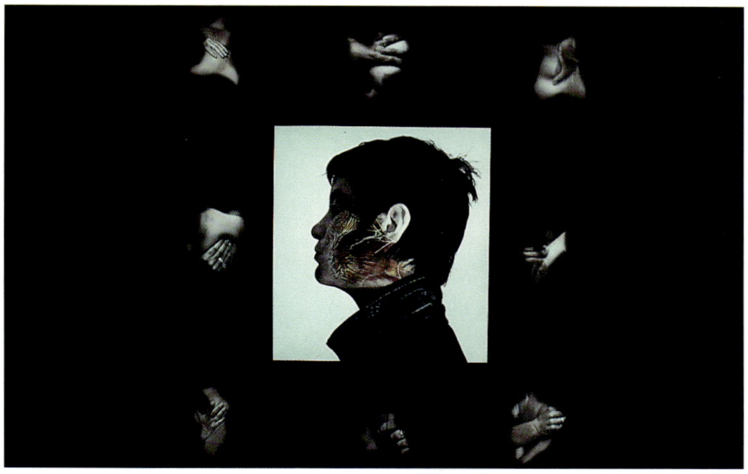

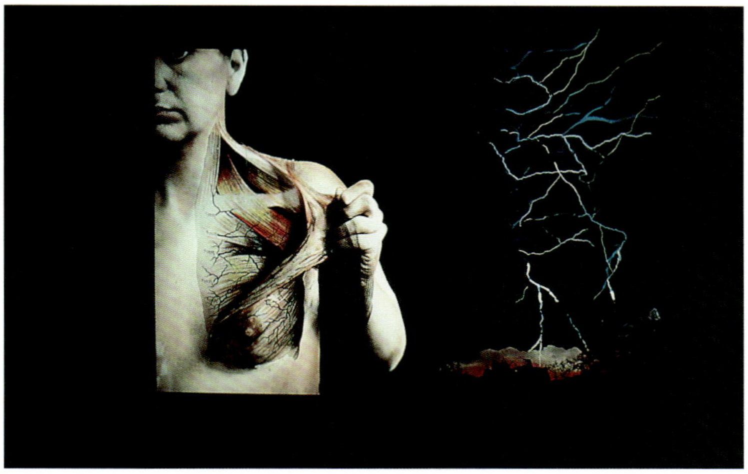

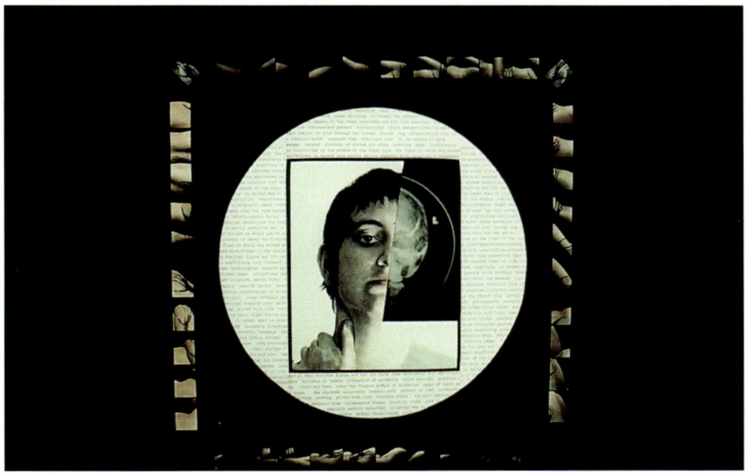

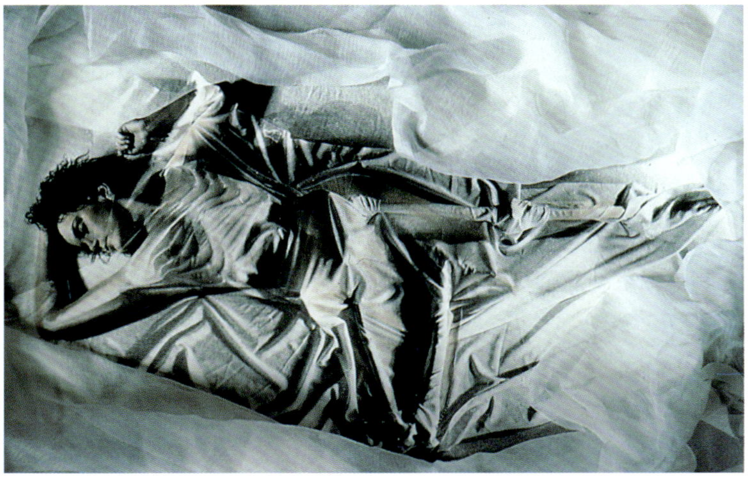

Roberta Graham, FROM ABOVE: L'Amour/La Mort, *1982/3 lightbox;* Whether the Storm?, *1983, lightbox;* Face to Face, *1982, lightbox;* Life Sighs in Sleep, *1983, sculpture*

When they returned to San Francisco, the professor took her under his wing. He taught her how to lead patients, sometimes by means of trance, through regressions, childhood and past-life. When he began to absent himself from sessions with patients, Georgina, in effect, took over his practice. At that time, she was suffering from allergy; her allergy doctor, who hadn't been able to help her, started sending her his patients when he saw that she had been able to cure herself.

Georgina sent me to Greg Schelkun.

My meeting with Greg signalled my entry into a school. A school in which I, who had done things like bodybuilding for most of my adult life, began to learn about my body. I want to introduce two of the people whom I met in this school:

Greg is a big, gentle man who looks like a Midwestern American farmer. Originally, he wanted to be an artist. Graduating from Dartmouth College in art history, he intended to go to Japan to study art. When, the year after he graduated from college, his mother, ill from a non-specific abdominal infection, asked him to accompany her to the Philippines to meet a healer, he had his first taste of what was to be his work. The healer cured both the mother's infection and the son's migraines. 'I couldn't deny what I was experiencing', Greg said to me.

Stubborn and still wanting to study art, Schelkun travelled to Japan. There, through a British artist, Rodney Cladwell, he realised that he wanted to work with Henry Moore. Schelkun wrote to Moore, then visited him in 1972. 'He was one of the most important people in my life', says Greg.

Moore told the ex-university student that he would be better off working on his own, but nonetheless, Moore would accept him as an apprentice. Since there was a waiting-list, there would be a year to a year-and-a-half delay.

'I was still intrigued by healers, wanted to make a film on them, so I went back to the Philippines. Two days before I was supposed to leave there, I met Placido. He was leaving 'on a mission' to the hills in three days. I asked if I could go with him. I didn't know why.'

Schelkun thought that he was going into the mountains for five days. Instead he was there for two years. Then, on the invitation of a patient, he went to northern California. Word-of-mouth about his skills soon brought him patients; he has now been practicing in northern California for 20 years and has been written up in various American magazines including *Time*.

I asked Greg what exactly he does.

'By seeing, listening, asking questions, I sense energies, how they're blocked. I want to make energy flow freely.'

'How do you do that?'

'It's up to your body . . . ' Greg hesitates. I know that he tells me only what he knows I can understand. He is always teaching. 'I'll tell you what this is all about.'

By 'this', I presume he means disease. Disease, which here is equivalent to life, for bodies are always changing, going through what we call 'disease'.

'It's about learning. We're alive so we can learn. Bodies are part of the challenge to learn about compassion. We learn from diseases; they are gifts.'

I ask Greg what health might be.

'Health is corporeal existence. We say 'good' health and 'bad' health, but we're only making up what 'good' and 'bad' are. We put value judgments on the body, whereas we should listen to the body.'

'You're teaching me the vocabulary of my body, aren't you?' I'm trying to understand what Greg has been doing to me – or is it with me? – all these months.

'That's the first step.'

'What's the next step?'

'To focus attention. That's how you get rid of symptoms.'

'Is there another step after that?'

'You have to want to be well. You have to learn what well is. That can take a lifetime or five lifetimes', he answers in his laconic way.

When I had first started working with Greg and Georgina, I had been confused about why I had gotten cancer. Three weeks later I saw the network of causation so clearly I wondered why I wasn't more disease-riddled. Georgina reminded me that if health is based on forgiveness, then I had to forgive myself. Again I inquire what health might be.

'The body remembers, especially traumas', Georgina thinks aloud, 'and holds these memories as scars, as wounds. Disease is when the body isn't in harmony, when there are areas of blockage in the body. So in order to lead someone to healing, for I work verbally rather than as Greg does, I roto-root the past. When a person goes through regression, childhood or past lives, that person is able to situate the trauma in the whole picture and so stop obsessing about it. For instance, take the blame off mommy and daddy; begin to see mommy and daddy as people in larger situations. All healing has to do with forgiveness. A healthy person is one who can say, "I no longer have scars from the past that will keep me from doing what I have to do today".'

'What is health?' I ask again.

'A body that's in harmony.'

'What about the mind?'

'The mind exists throughout the body.'

During the period in San Francisco when I worked with Greg, Georgina, and Ellie Wood, a former student of Greg's and now a healer who taught me how, through visualisation, to examine my major energy centres, I fell in love with someone who's English. At the end of July, when my healers told me that I was on my way to health, I moved to London. where I began working with Ken Lloyd, a Chinese herbalist, Beverly Katz, a cranial therapist, and Stephen Russell, 'The Barefoot Doctor'.

Russell is certainly the most colorful of this group. Deciding after his first term at university that he would rather be in the world, Stephen started importing Italian fashion into London.

When he met RD Laing, the Scottish psychiatrist and author of *The Divided Self* (1960), he told him that he was going to be his student.

'I don't take private students', replied Laing.

'You're going to be my teacher!' Stephen exclaimed.

Ronny laughed and took him on.

They met twice a week, once for therapy Laing-style and once for discussion.

Then it was 1979 and Russia invaded Afghanistan. Convinced, perhaps strangely, that a nuclear disaster was about to end the world, Russell left England, his studies with RD Laing, and his business, in order to take his family to the United States. There he intended to meet up with Thomas Benyaka, the Keeper of the Prophecy for the Hopi Indians, whom Frank Waters had mentioned in *The Book of the Hopi*.

After journeying over 11,000 miles, Russell found the man for whom he was searching. 'I asked him if the world was going to end and Thomas Benyaka said, "No, our prayers have taken away that disaster". Then he said, "You have to go away and heal yourself before you can heal other people". At that time, I had never thought about the word "healer" much less about me being a healer. I just wanted a way out. This man told me who I was going to be and then he told me that I had to go away.' Stephen travelled to Taos, New Mexico, and went to work in earnest. He studied with Joe Sawasso, an American-Indian healer, and Han To, a Chinese herbalist from San Francisco; not to mention Morgan Shannon, a well-known Reichian therapist.

Though trained in different disciplines, Schelkun, Ritchie, and Russell speak about disease in the same manner. 'Disease comes from blockages of energy', says Russell. 'I heal someone either by getting rid of these blockages or finding bypasses so the energy can again move freely.'

'How do you do this?' Though Stephen has worked on me, or with me, many times, I cannot articulate what he does.

'Every person has two selves: a local and a transcendental. The transcendental self is the one who watches. I help a person get there, for if he can, he can see or find how to heal himself.'

I remember Greg talking about disease as a school, saying that he is teaching me how to heal myself.

'I want to say one other thing', mentions Stephen. 'If you look from the point of view of the whole, the whole is synergy. The whole universe is always moving, healing itself. It all comes down to faith.'

Does it? I have listened so closely to my healers' words that, perhaps, I have forgotten to ask myself how I perceive healing. How do I understand what has been happening to me or in me?

When I walked out of the surgeon's office and didn't know where to go, I asked myself what I could know. Did I have anything in myself, in my life, that could help me know and so deal with cancer?

My answer was: It takes strength to know. Where, then, is my strength? Answer: in my work, my writing. What are the tools of my writing? Imagination and will.

Imagination allowed me to envisage something other than the surgeon's doom tones; will actualised that which I imagined. I was able to make a leap of faith and knock on Georgina's door, then on Greg's and Ellie's.

I am discussing healing. Greg said, 'First, one must want to be well.'

What was this 'leap of faith'? When I asked Georgina for help in fighting cancer, she said with certainty, 'You will create your own healing'. I took it for granted, as I think many of us do, that doctors are people who know how to get rid of disease, that one turns to a doctor in order to be cured of physical affliction. Though I no longer trusted conventional doctors, I trusted myself less. How could I heal myself of a disease as serious as cancer? I thought Georgina mad. My 'leap of faith' was to believe that I could be healed and that I could heal myself. That I could heal without having to trust someone else with my life as I had done with the surgeon; that I could be responsible for myself. Georgina proved to be right. I am the one who must heal in order to be healed.

In order to heal, I had to question and even abandon certain basic medical paradigms. A doctor, for me now, is not someone who cures, for, as Greg says, no one cures life. Rather, a doctor is a facilitator and teacher and the doctor-patient relationship a spiritual, as well as a medical, one. The most radical paradigm shift, however, concerns the body. Since the healing work that I did and am doing and that was done to me was mental and spiritual as well as physical, I experienced my body either as having several aspects or as several connected bodies, as soul, as energetic or emotional, and as physical. I saw that the organs, the tissues and the cells physically remember all that once took place to and within the body. I experienced the body as an entity separate from all other entities only in its physical aspect.

A final paradigm shift concerns the experience of 'going to the doctor' or of healing. Once, having left the surgeon's office or the realms of conventional medicine, I had grappled with, then passed through the worst of fear – fears founded in birth and in childhood as well as in considerations of death – I felt only intellectual excitement and joy.

When I met Greg and entered the school of the body, I entered the school of the self. I thought I was, unwillingly, confronting cancer; I was confronting myself. I no longer have cancer. To heal in oneself is to begin to heal the self that is always whole.

I have written down some of what happened to me in the past nine months, though, as yet I understand little.

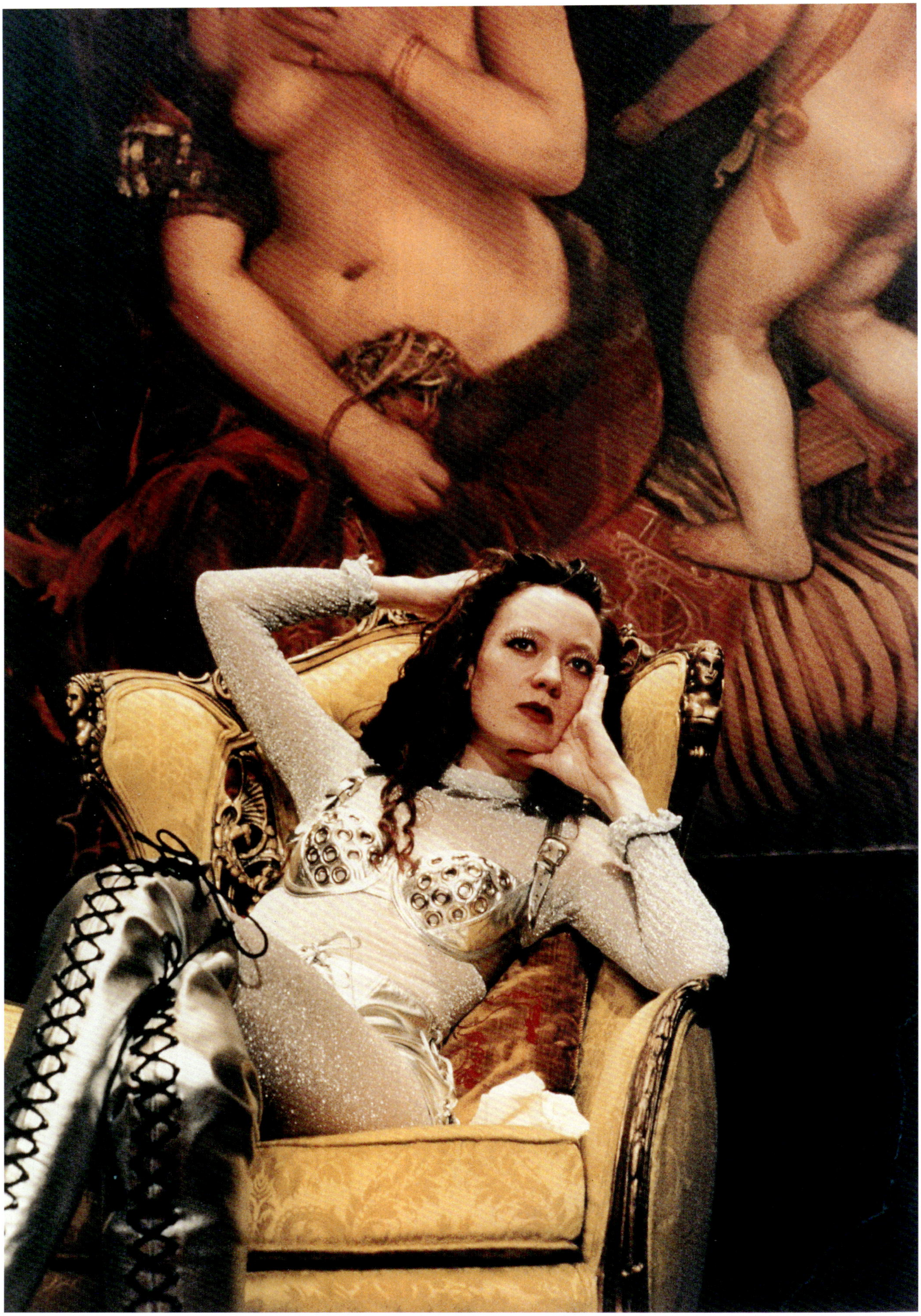

Nick Broomfield, Fetishes *1996 (Mistress Raven at Pandora's Box, New York) still from documentary*

UNREASON TRANSFORMED INTO THE DELIRIUM OF THE HEART[1]

Nick Broomfield's Fetishes *documentary*

JULIE WEIR

'Exhibitionist display. Voyeuristic pleasure. Experimentation with personal identity. The visualisation of erotic visions'[2] – fetishism can be viewed in a number of ways. Its rituals and activities have existed in human culture for centuries, with the investing of special, even sexual, powers into certain objects or materials. Religious history is full of flagellation, confinement and punishment, often undertaken in order to achieve a heightened state of communication with one's 'god'; and many respected historical characters indulged fetishes of varying degrees – from Mark Anthony and Claudius to Mozart and Napoleon. Today, fetishism in all its forms – bondage, piercing, rubberism, mummification, asphyxiation, infantilism etc – is stigmatised as a socially unacceptable form of behaviour, even though there is a great deal of curiosity surrounding its practice.

Nick Broomfield's documentary *Fetishes,* (1996), and the photographs of Nicholas Sinclair, Housk Randall, Grace Lau, Richard Kern and Doris Kloster, seek to demystify forms of fetishism, moving previously ghettoised obsessions into the public and mass-media focus, and incorporating them into fashion, art and language.

Some audiences may regard *Fetishes* as a showcase for abnormal sexual behaviour. However, since 14 per cent of men and 11 per cent of women (according to the Janus report), have experienced some form of fetishism, and fetish-related material generates a multi-million pound international industry, the practice cannot simply be dismissed as 'weirdness'. The film deals with an area of human sexuality that is unusual but not uncommon, and Broomfield claims that making it was a journey of discovery that has helped him to understand his own sexuality and relationships. His aim was to offer people information on a subject about which they are largely ignorant and which has previously been approached in a sensationalist manner. As such, it is a serious, non-gratuitous piece of documentary work that may still shock the easily offended.

Inevitably, the film, made in America and due to be screened in British cinemas in Autumn 1997 prior to its video release, has generated controversy and calls for censorship. Broomfield believes that it is highly doubtful that a film of this type would have been made in Britain. It remains to be understood why Beeban Kidron's equally controversial *Hookers, Hustlers, Pimps and Their Johns* has already been shown on Channel 4, yet schedulers remain too squeamish to deal with *Fetishes.* Kidron's film is a gratuitous examination of the seedier side of the sex industry, while *Fetishes* portrays its subjects as real people with certain needs, not as perverts.

Fetishism and S&M is an economically viable industry with a thriving trade in clothing, magazines and clubs, but a great deal of naivety surrounds this piece of the human physical and emotional jigsaw. Sexuality involves more than an instinctual response to a programmed stimulus and is a product of both history and nature. It is learned, not constructed in the sense of acquiring a set of rules. Hence, fetishism demonstrates that the sexual instinct and the sexual object occupy an extremely unstable relationship. Indeed, Alfred Binet (1857-1911) believed fetishism was a matter of association – a conditioned reflex similar to that created in dogs by Ivan Pavlov – amplified by robust and intense emotions that free the imagination 'to choose what will produce sexual satisfaction'.[3]

The commonly held view amongst psychologists is that fetishism develops in an individual at an early age – according to Freud it is a variety of the primary phase of the sexual function. Adult fetishistic behaviour involves the eroticisation of traumas experienced in childhood, which are reversed and manipulated into a happy ending. Broomfield sees this as 'the eroticisation of your worst nightmare' and believes that 'to be the author of your own nightmare must be a truly liberating experience'. Fetishism also taps into childhood memories, conjured up by objects, places or occasions that provoke simultaneous feelings of fear and attachment. This is a combination of respect, desire and terror that we take as the psychological sign of the sacred. The term 'fetish' is therefore often used to refer to anything 'irrationally' worshipped.

To Broomfield, fetishisation involves the investment of objects or clothing with sexual powers – an exaggeration of what is already there. ('Everyday dressing is sexual, but fetishism is a eulogy to it.') Freud suggested that a fetish was an item that appeared alongside the 'normal sexual aim and object and ousted it completely.'[4] Through these attitudes comes the much contested belief that fetishism may be 'a kind of compulsion, a combination of unusual brain wiring and aberrant conditioning'[5] – part rebellion, part psychosis.

Pandora's Box, the setting for *Fetishes* , is a lavish, upmarket S&M parlour in New York. No straight sex is offered here, though a multitude of different fetishes are catered for both through carefully designed rooms (the medical room, the dungeon etc) and the various outfits and uniforms worn by the dominatrixes. The wealthy clientele are submissives who wish to be punished for some fictitious wrongdoing. Hence, the core relationships seen in the film are dominant/submissive or mistress/servant. The most common services paid for by males in heterosexual S&M are the extravagantly theatrical displays of submission seen at Pandora's Box.

It is a complex psychological and physical role reversal: in most commercial bondage and disciplinarian scenarios males are the slaves to female dominants. Scenarios are strictly scripted in all aspects and a 'save' word is issued to call a halt to the proceedings. Hence, it is the submissive 'bottom', though often extremely emotionally needy, who actually has control over what occurs. In this temporary relationship the dominatrix must not go beyond the accepted rules. So, it seems that passivity is covert control. The 'top', virtually impotent, is continually shaping the game to the submissive's limits.

It is the pain element of sadomasochism that carries the greatest stigma, because 'physical violence, particularly to someone of the opposite gender is taboo and productive of intense guilt. To strike is sin; to be struck is guiltless or virtuous in a martyrdom sense'.[6] But at Pandora's Box the giving and receiving of pain is part of an interactive system of concern and respect, and it is 'the illusion of violence rather than the violence itself that is arousing to both sadists and masochists. At the very core of sadomasochism is not pain but the idea of control – dominance and submission.'[7] Pain per se, is not attractive to the masochist unless it occurs in an arranged situation. Mistress Raven, the head dominatrix at Pandora's Box, quotes De Sade's observation that there is a very thin line between pleasure and pain. The infliction of pain is merely a method of achieving the power exchange in these scenarios. Definitions of these activities that focus exclusively on pain miss the essence of sadomasochism, which is that the art of S&M is theatre – a delicious simulation of harm and high risk. This symbolism structures all interactions in this realm.

It is important to acknowledge that all participants are willing, though roles are strongly differentiated. These are expressed by postural and verbal constructions associated with power, control and authority. If the dominator's mask evokes the executioner, the slave's leather hood implies the victim. The submissive partner will either be naked or wearing restrictive clothing associated with servitude. They may also be ritually branded or pierced to indicate permanent ownership by their master or mistress. These activities might be seen as ridiculous or barbaric by many, but there is some innate quality in all of us that responds to violence – from the crowds that flocked to public executions to the audience at the big fight. Some degree of aggression and submission is evident in most, if not all, sexual relationships, even if only verbal.

What is important in S&M fantasy is not the object of desire so much as the creation of a certain scenario where particular wishes are presented. The development of an identity in this 'world' is part of an interactive process, beginning with the recognition of certain feelings and culminating in learning to construct and control S&M situations. The issues of limits, fantasy and control are all intertwined and directly related to trust. At Pandora's Box, clients fill in a questionnaire regarding fantasies, clothing, fetishes and pain thresholds, enabling the mistresses to create a 'pleasing' scenario for the client. Typically, rules are made to be broken and the punishment that follows tends to be theatrically extreme. Sadomasochism shares this theatrical iconography of punishment and expiation with Christianity and similarly performs the paradox of redemptive suffering through the punishment of the flesh.

Some of the scenarios in *Fetishes* take a socio-political twist – for instance, a Jewish client who has Nazi fantasies and an Afro-Caribbean client who enjoys plantation and slave scenarios. Broomfield believes this to be a political statement and a reflection of the society in which we live, claiming that a great deal of unresolved problems reach closure for many clients at Pandora's Box, who treat it as a different world.

An investment broker with a corporal punishment fetish says he has an insatiable drive that must be fulfilled; 'Slave Jeff', dressed in a PVC maid's outfit, claims 'it is a lifetime passion' to serve the mistresses; the 'human ashtray' asserts that it is a pleasure to be in their presence; and a rubber fetishist claims that he must visit Pandora's Box because he cannot share his fantasies with his wife. Jessie and Maria are rare female clients. Here, the activity is similar to other scenarios but the symbolism is different, with one woman dominating another rather than the opposite sex. This offers female submissives an opportunity to transcend social norms that restrict their behaviour, in the safety of a controlled fantasy. Broomfield notes that the mistresses seem to enjoy these sessions the most.

The mistresses of Pandora's Box represent the opposite of the traditional view of women sanctioned by conservative society – they are neither passive, demure, nurturing nor maternal. They make a positive statement towards powerful, Amazonian femininity, holding their worshipping clients on an invisible leash. However, only Mistress Raven seems to be consistently convincing in this role. ('A genuine dominatrix is just a little scary and she gets your attention when she whispers just as easily as when she shouts.'[8]) She never lets Broomfield interrupt her when she is speaking, and always maintains her powerful demeanour – unlike the other mistresses, who tend to turn into giggling schoolgirls when interviewed. Broomfield believes that his film portrays the mistresses as likeable, real people, emphasising the positive side of what they do – not as administrators of pain but as 'technicians of pleasure'[9]. Mistress Catherine believes being a dominatrix has made her stronger in outside relationships, both sexually and in friendship. But Broomfield also deals with many of their misgivings and problems. Mistress Delilah, for example, worries about what people will think of the film: though her participation in this particular kind of work has taught her to be tolerant, not everyone will have the same approach. Mistress Natasha says she loves men, but understands that this may sound confusing when she 'beats them up' for a living. She claims that you have to like men to work in the sex industry because it is here that you see their worst side. In commercial S&M the dominator acts as an official, if forbidden, witness to private anguish and confused desires.

It is Mistress Raven's relationship with her work that is the most interesting. She is a famous 'lifestyle' dominatrix – her work extends into her personal life, and her husband is also heavily involved in the scene. She runs Pandora's Box, but has ceased to conduct sessions with clients because she finds these needy people emotionally and physically draining. Nevertheless, Mistress Raven is still very much the face of domination at the parlour and extols the virtues of the services offered there, claiming that the experience is a perpetual journey of knowledge that makes life more complete and complex.

Why do people engage in sadomasochistic activities? One of the most popular explanations is that it offers escape from modern pressures and responsibilities. Another, that it is to do with the stimulus of the forbidden, where the imagination acts as a substitute for reality. RF Baumeister believed that pain acts as a 'technique for removing higher level self-awareness, while promoting a low-level awareness of self as physical object. Pain brings self awareness down from symbolic identity to physical body.'[10] Sadomasochism may thus be approached in a similar way to the feats of fakirs and practitioners of Eastern religions, involving the same mechanisms of dissociation as can be found in hypnotic states. 'Through self-abasement the spirit finds release in an ecstasy of abandonment.'[11] Furthermore, it has been shown that the infliction of pain triggers the production of endorphins, which generate a feeling of euphoria. Seen in this light S&M goes far beyond the exotic or the idea of a mere 'holiday from life's responsibilities.'[12]

The S&M subculture is dependent upon certain meanings that are culturally produced, learned and reinforced by its participants. Individuals frame their behaviour in certain ways, by the use of social definitions that give it a specific contextual meaning. Those involved in this world understand what is happening in the various 'scenes', but an outsider who does not comprehend the conventions may be confused, offended or upset by the blatant display of exaggerated violence, abuse, racism and anti-semitism in *Fetishes*. Failure to view S&M behaviour within its social context risks misinterpreting what is going on. Within the realms of S&M nothing is seen as abnormal. Both males and females can be dominant and submissive and people may act out any fantasy they have. Hence, it could be claimed that S&M provides is a higher plane of existence – a world without prejudice.

Notes

1 M Foucault, *Madness and Civilisation: A History of Insanity in the Age of Reason*, R Howard (trans), Tavistock Publicatons, London, 1965.

2 T Polhemus, *Rituals of Love*, Macmillan, London, 1994, p64.

3 Colin Wilson, *The Misfits*, Grafton Books, London, 1989, p175.

4 Bill Thompson, *Sadomasochism*, Cassell, London, 1994, p92.

5 C Gosselin and G Wilson, *Sexual Variations: Fetishism, Sadomasochism and Transvestism*, Faber & Faber, London, 1980, pp153-156.

6 P Gebhard, 'Fetishism and Sadomasochsim', in JH Mosserman (ed), *Dynamics of Deviant Sexuality*, Grune and Stretton, New York, 1969, p42.

7 Weinberg and Kamel 'An Introduction to the Study of Sadomasochism', 1983, in Thomas S Weinberg (ed), *S&M: Studies in Dominance and Submission, Prometheus Books*, New York, 1995, p20.

8 Pat Califa, 'The Power Exchange', in T Woodward (ed), *The Best of Skin Two*, Masquerade Books, New York, 1993, p59.

9 Bill Thompson, *Sadomasochism*, Cassell, London, 1994, p147.

10 RF Baumeister, 'Masochism as an Escape from Self', in *Journal of Sex Research 25*, p39.

11 A McClintock, 'Made to Order – Commerical S/M and Gender Power' in Gibson and Gibson (eds) *Dirty Looks: Women Pornography and Power*, BFI, London, 1994, p222.

12 T Polhemus, *Rituals of Love*, p84.

FERNANDO ARIAS

What is he Bid?

CATHERINE WILLIAMS

The work of Colombian-born artist Fernando Arias incorporates South American religious and political imagery, the body as material, and sophisticated computer and video technology. A number of years spent in his father's hospital workplace offered an ideal creative environment and access to materials from human umblilical cords to computerised images of foetal scans and viral-infected blood cells. Exploring his country's cultural ability to amalgamate both ancient beliefs and modern ideas, and using his own fluids and flesh, Arias confronts the body as a highly politicised space.

Seropositivo, which will tour the UK in 1998, is a larger-than-life photographic portrait of the artist, naked and exposed, his limbs outstretched, placed onto the gallery floor and covered with a transparent layer of thousands of anonymous blood samples. Illuminated by UV light the artist's body appears to be suspended in air. The resulting sanctuary-like atmosphere creates a space for considered reflection on our ethical responsibilities. The random arrangement of anonymous blood samples, some infected, some healthy symbolises the fragile link between life and death and represents a complex social framework. Technology, which is globally applauded for developing medical research, is exposed for its inability to protect us from disease.

The massed, floating cells that appear in many of Arias's AIDS-related installation works are also symbolic of the denial of a fertilising destiny for millions of potentially life-producing sperm as they journey through ejaculatory fluids from one man to another, swimming against the views of the Catholic Church. This lack of procreative goal reinforces the Church's damning of the homosexual act, which it also justifies by isolating the cause of the HIV virus to only one form of exposure – the mixing of blood and semen. This conveniently targets the homosexual community and the economically and educationally deprived 'third world' countries – the social groups to which Arias belongs. As a lapsed Catholic, Arias has often used religious iconography and other symbols to challenge the Church's failure to accept his sexuality.

Arias's politically confrontational installation works have caused great controversy in his country. In October 1996, the director of a major museum in Bogota refused to sponsor his new work for the Bienale de Bogota on the grounds that Colombia could not afford to fund cultural and educational projects. Arias's response was to secure the inclusion of a single, highly provocative work. In a glass cube containing four litres of his own dried blood Arias reproduced an extract from the director's letter employing lines of cocaine and the synthetic powder used for cutting it. The poignancy of this statement lies in the orgainised drug-dealing that funded the recent Colombian presidential campaign and involved the director's son.

A further series of sculptural objects made from glass, metal and blood resulted in Arias's only commercial sale to date, to a Canadian collector who, finding that there was no visible signature on the work, insisted that Arias scratch it onto the surface. This experience, along with an invitation by a curator of Arco 97 to produce multiples, led to his current, highly self-intrusive work, which uses the body for political comment and explores the commerical art fair environment and its craving for 'signatures'.

For the ironic project *What is he Bid?* Arias has tattooed his signature onto his body and is auctioning the 3 x 9cm fragment of skin which will be surgically removed and presented in a formaldehyde-filled glass case. Arias' self-designed brochure carries the logo *Crisi's*, an ironic play on Christie's (the international auction house) and the poignancy of the word 'crisis'. Distributed as a multiple throughout the Madrid art fair and other exhibitions, the brochure includes a colour photograph of the 'art object' and a simple form on which an individual or organisation can nominate a bidding price. Whoever mails the highest bid will become the owner of the tattoo.

So what is Arias offering? Not a recognised signature, since it has only once (and then unconvincingly) appeared as a definable mark on his work. His reputation as a visual artist does not merit such an egocentric statement. So is this a comment on the art world forcing the artist to degrees of self-mutilation? Does the desire of the art buyer demand such proximity to the body of the maker; the obsessive fan tearing at the flesh of the idol? What is being offered is literally a piece of the creator; a slice of the artist's culture for a Western market that has been historically preoccupied with obtaining and repackaging icons of ethnic authenticity or exoticism at his country's economic expense. Arias is tendering the ultimate cultural purchase, a wafer of skin floating in fluid like the ceremonial offering of the symbolic body and blood of Christ – complete with signature.

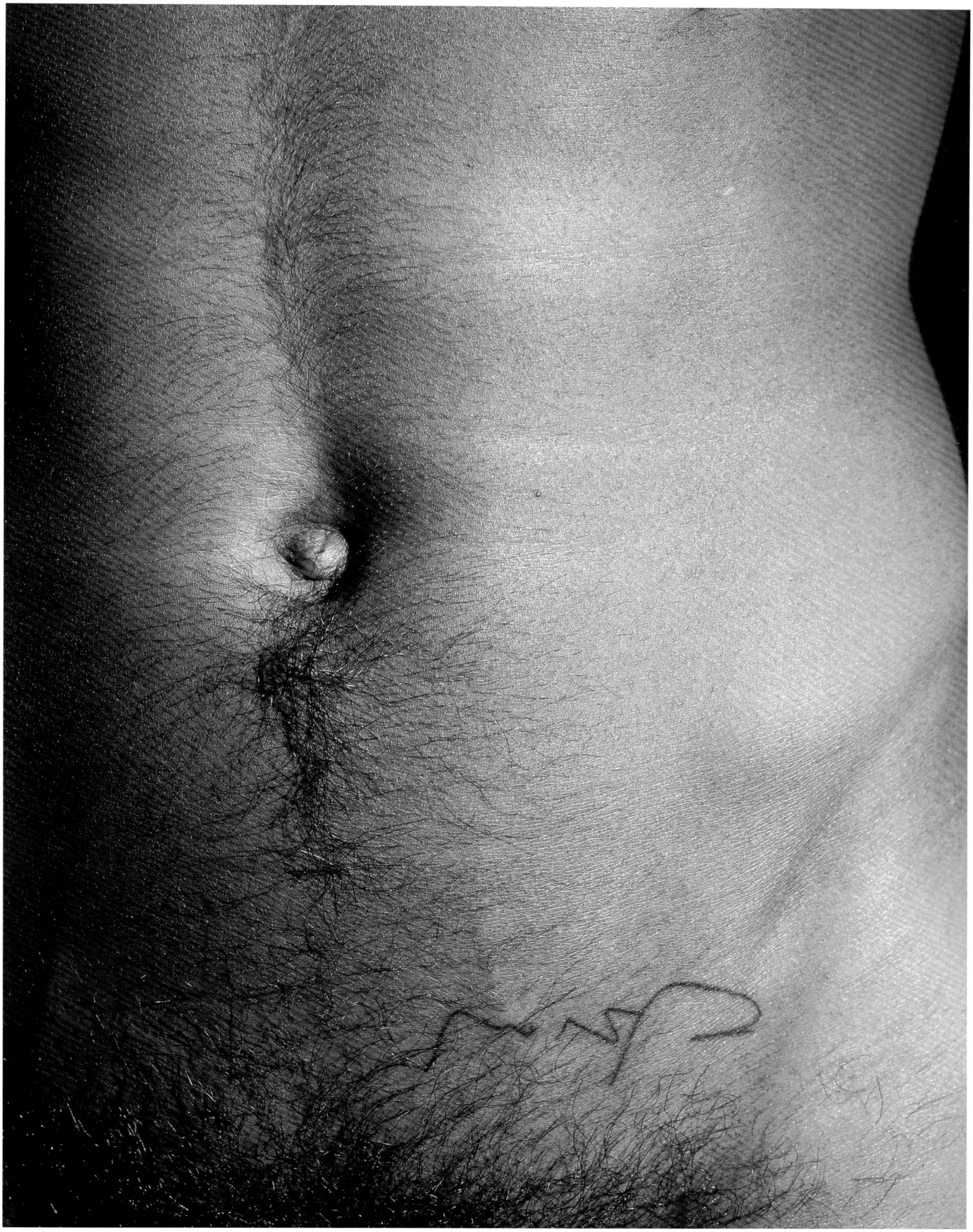

Fernando Arias, What is he Bid? (¿Quién da Mas?), *1997, tattoo (photo: Julian Velazquez)*

PREDRAG PAJDIC

The Body as Costume

Predrag Pajdic is a unique artist and costumier interested in designing for the interior, as well as the exterior, of the human body. Trained in electronic engineering and fine art in addition to fashion, Pajdic combines skills from all three in an original exploration of the human form. In his work, he pioneers a new designer-led reconstruction of the body as pret-à-porter.

Pajdic's lasting fascination with anatomy reflects his view of the human body as the most complex and absorbing, cybernetic organism. Not merely concerned with the body's aesthetic form, nor subscribing to the Victorian obsession with the grotesque or 'freakish', Pajdic encourages respect for the body by understanding it as part of the natural world, and therefore as a thing of beauty. He likens it to a chemist's laboratory containing thousands of tubes and vessels filled with their own compounds and performing their own function. The removal of any one of these components unbalances the function of the body as an integrated system, in the same way that changing a single element of a chemical compound may convert it into an inert or explosive substance.

In contrast to Leonardo da Vinci and the Renaissance artists who tried to rationalise phenomena that were not understood at the time, Predrag Pajdic experiments with medical and scientific truths and playfully re-examines them. Taking existing body components and replacing them with invented ones, he reassembles conventional arrangements into something more offbeat and unpredictable. This ludic ideal serves as the basis for his unique designs.

His garments are made of unusual materials such as metal, paper, plastic and rubber, which produce a translucent, protective barrier around the body, suggestive of exfoliated reptile skins. These delicate, yet tensile robes, seem to have been naturally manufactured by the body itself. The body that is dressed in Predrag Pajdic's clothes is therefore capable of metamorphosis, changing its skin to suit the occasion or the personality that currently inhabits it. While his metal mesh garments stress the transformation of outer layers, his transparent plastic garments act as thin lenses that provide a distorted view of the body, giving the impression of an alternative anatomy.

These protean shapes and textures, incorporating 'intelligent' fabrics and nanotechnology, echo the autonomous skins described in Cyberpunk novels, such as Pat Caddigan's *Fools*. But Pajdic does not stop at the surface of the body to create this chameleon screen. In his garments, he plays on the cliché 'You are what you wear', extending this to the inner workings of the body and its secrets, and examining our fear of our own inner identities as a problem in costume design.

These concerns have inspired him to make use of medical students' dissection manuals and found objects such as bones, washed up on the banks of the River Thames near his home. He incorporates these unorthodox fragments into his designs and exhibits them as autonomous art works, inviting his audience to look at the interior of the body as well as the exterior, without resorting to shocking, gory imagery. Unlike artists such as Orlan, whose work is obscured by blood and the expectation of pain, Pajdic gives audiences the opportunity to gaze upon the open flesh of his redesigned body and imagine alternative anatomies that they may wish to wear or inhabit. He tries to comfort the viewer by demonstrating that even when body parts are reorganised, this can be done in a playful and respectful way that produces something beautiful: a parallel view of nature. *(Rachel Armstrong)*

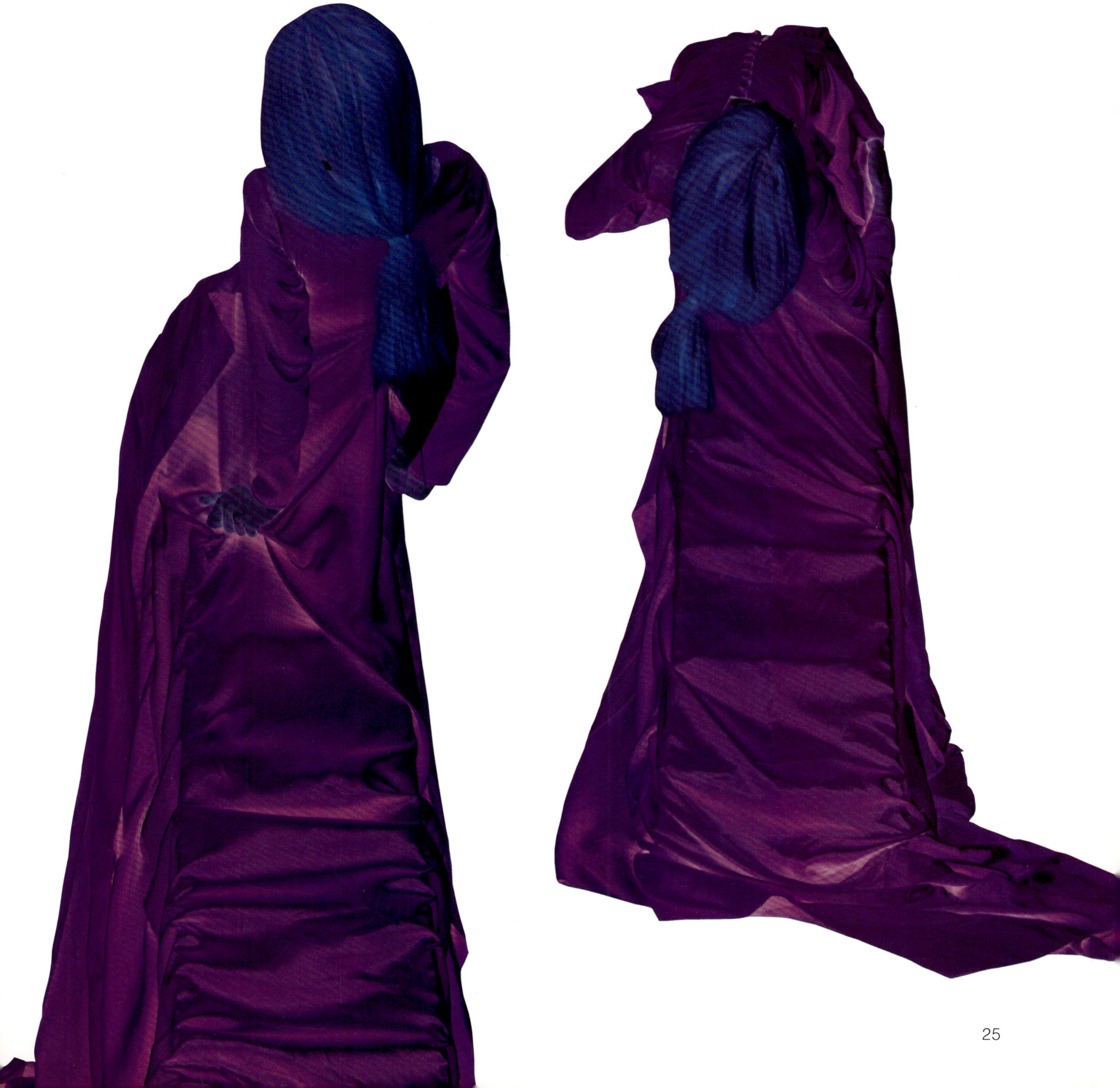

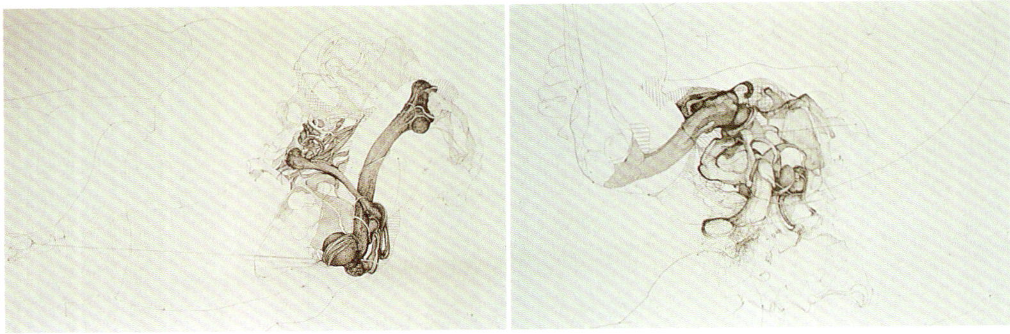

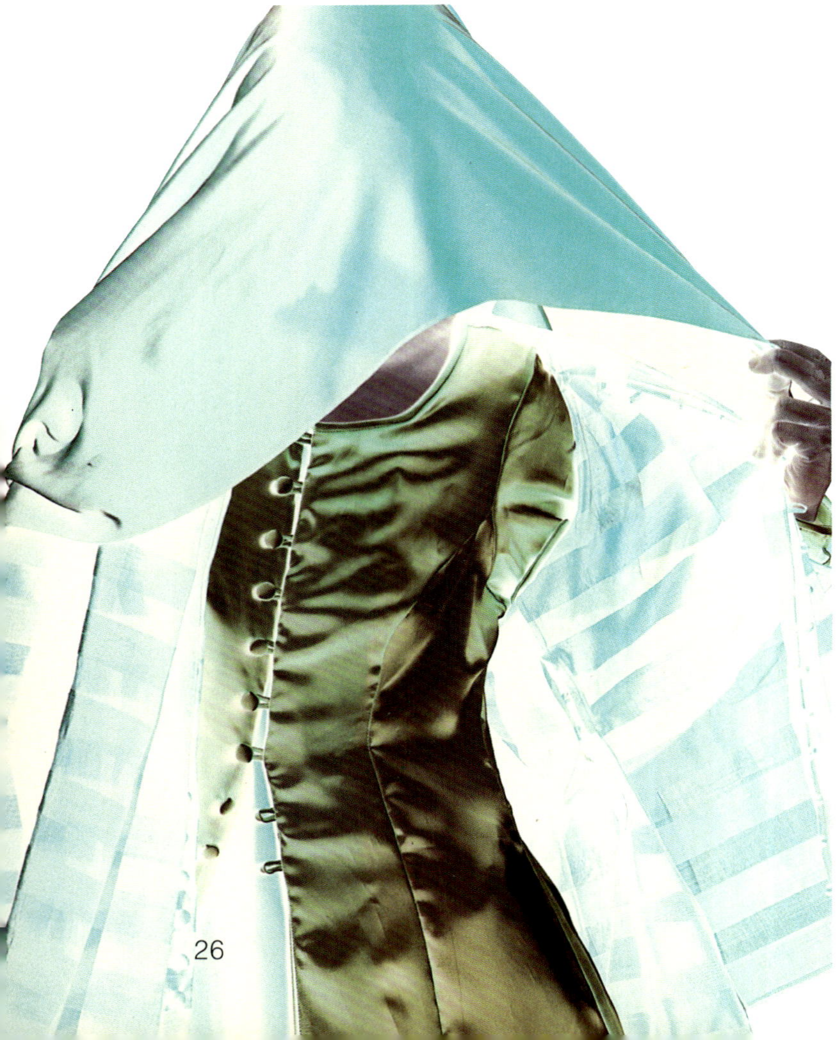

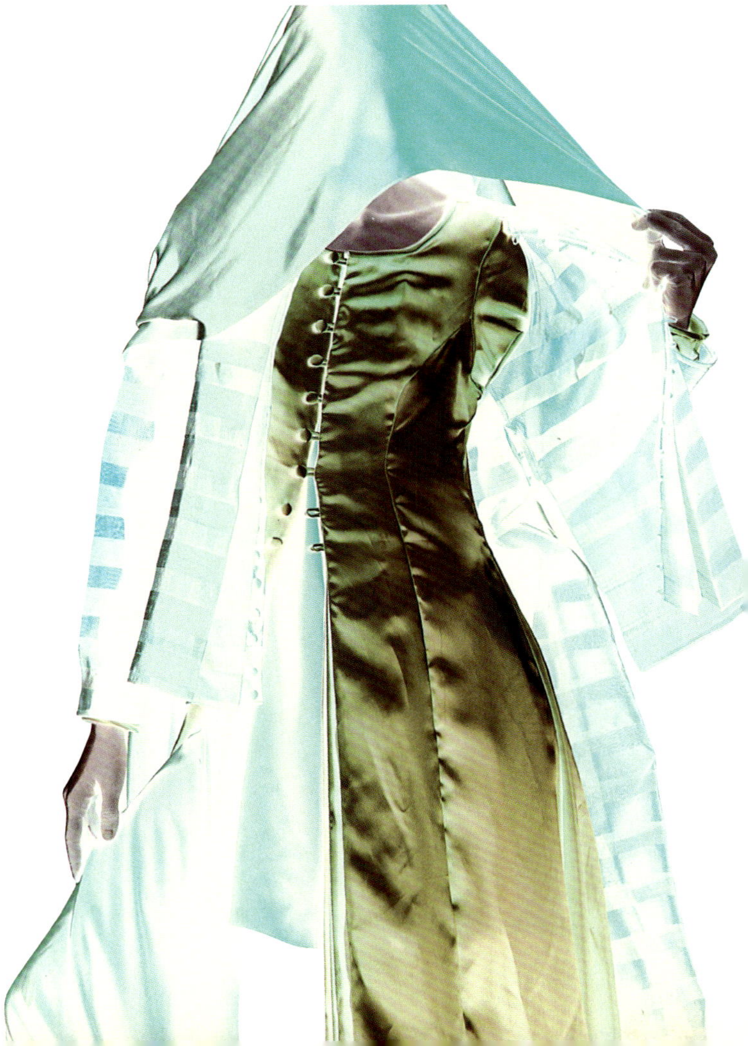

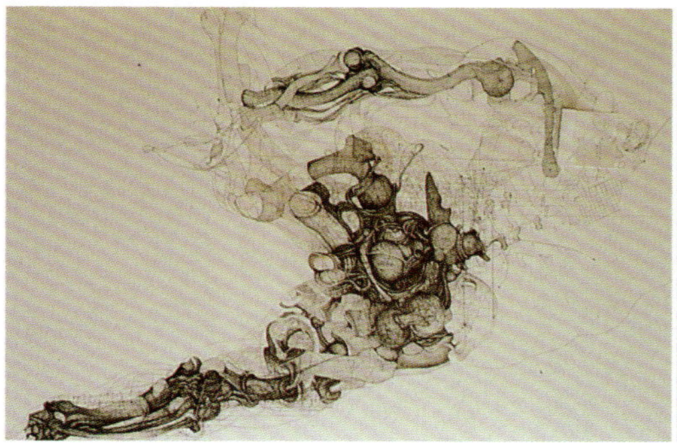

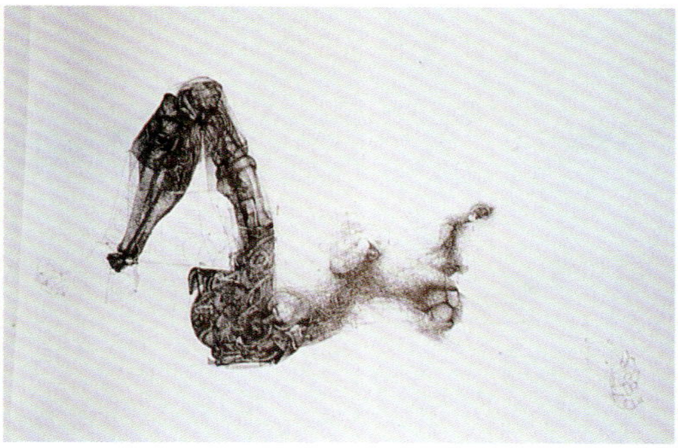

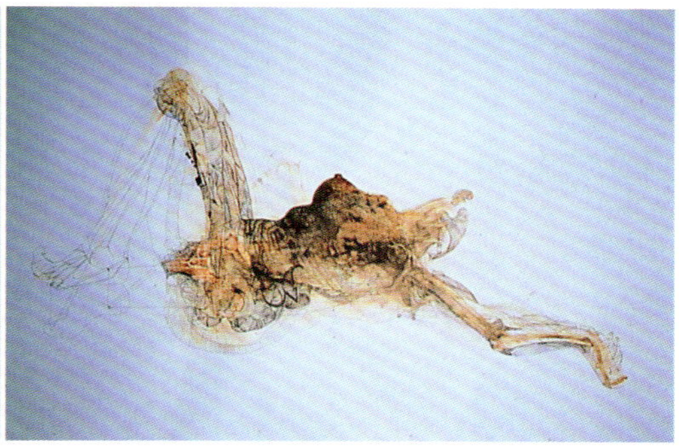

FROM L to R: Figure I, *pen and ink on canvas, 200 x 130cm;* Figure II, *pen and ink on canvas, 200 x 130cm;* Figure III, *pen, ink and bleach on canvas, 200 x 130cm;* Anatomical 1, *pen and ink on paper, 225 x 122cm;* Anatomical 3, *pen, ink and bleach on paper, 225 x 122cm, all works 1996*

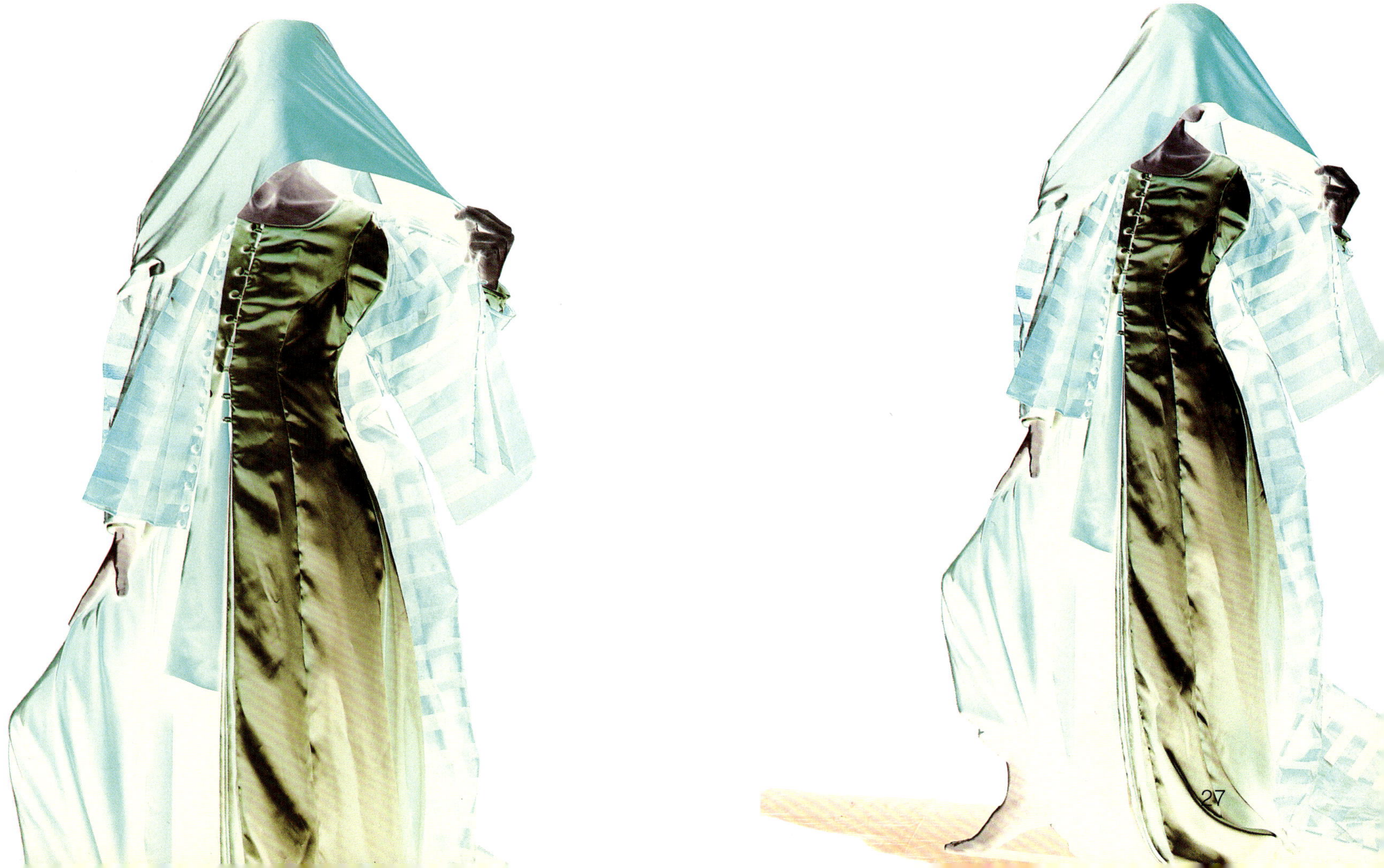

ORLAN

Carnal Art

As the first artist to use cosmetic surgery to alter her body, Orlan is at the forefront of a shift in social attitudes that will define the 21st century. She has literally given her body to art and uses surgery as a medium to confront the values of aesthetic surgery itself. She does not attempt to look better or younger, but to alter the process of cosmetic surgery in order to challenge and change its objectives.

She began her series of operations/performances in 1990 at the Edge Festival in Newcastle upon Tyne, with *Image/New Images . . . The Reincarnation of Saint Orlan*, and is scheduled to finish them in 1999 with the creation of a very large nose. So far she has carried out nine of these elaborate art works, the last, *Omnipresence*, 1992, being a collaboration with the surgeon Dr Marjorie Cramer, who inserted the largest cheek implants possible for Orlan's anatomy at each temple to create two bumps, referred to by the media as 'horns'. After surgery, Orlan literally wears her artwork every day. When the operations are finished she will approach an advertising company to come up with a new name, and will contract a lawyer to petition the French Republic to accept her new identity. This will be a series of performances testing the process of self-determination in society and challenging the legislation that governs one's identity.

At the beginning of each operation/performance she reads aloud from *La Robe* by Eugénie Lemoine-Luccioni, a Lacanian psychoanalyst: 'Skin is deceiving . . . in life one only has one's skin . . . there is a bad exchange in human relations because one never is what one has . . . I never have the skin of what I am. There is no exception to the rule because I am never what I have.'

Orlan believes that we now have the means to close this gap with the help of surgery, and that it is therefore possible to match the internal image with the external one. She sets out to construct an alternative self-portrait using computer technology, and to challenge the dominant notions of beauty and identity in collaboration with the surgeon. Remaining fully conscious during the process of cosmetic surgery she is witness to her own metamorphosis.

Each operation has its own style, ranging from the carnivalesque to high technology. Her series of operation/performances have been based on the following themes: Carnal Art, Identity Change, Initiation Ritual, This is my Body . . . This is my Software, Successful Operation(s), Body/Status, Identity/Alterity.

During the operation/performance she reads psychoanalytic texts related to each specific theme for as long as possible, even when the surgeon is working on her face. Throughout the procedure the operating theatre becomes her studio, in which she produces autonomous images, makes a film, a video, photographs, objects and reliquaries that are later exhibited in galleries. Thus, the once sleeping patient awakens and participates in the theatre.

Orlan's 'live' broadcasts are startling, even horrific to a general audience, provoking questions about our expectations of

Omnipresence, *1992*

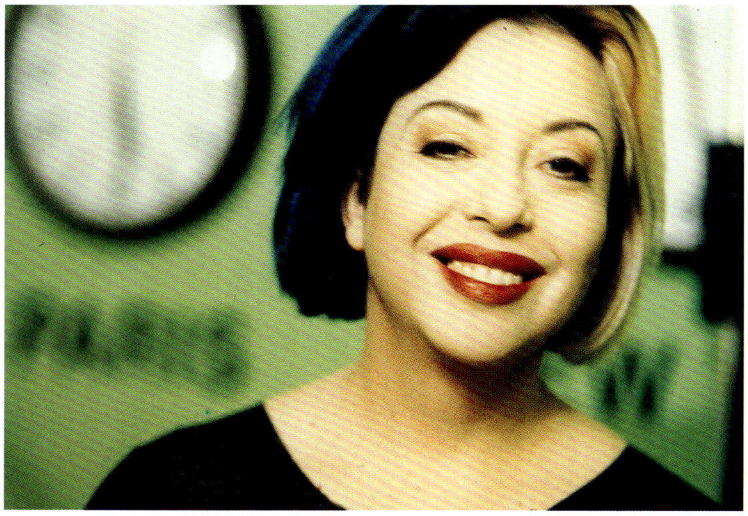

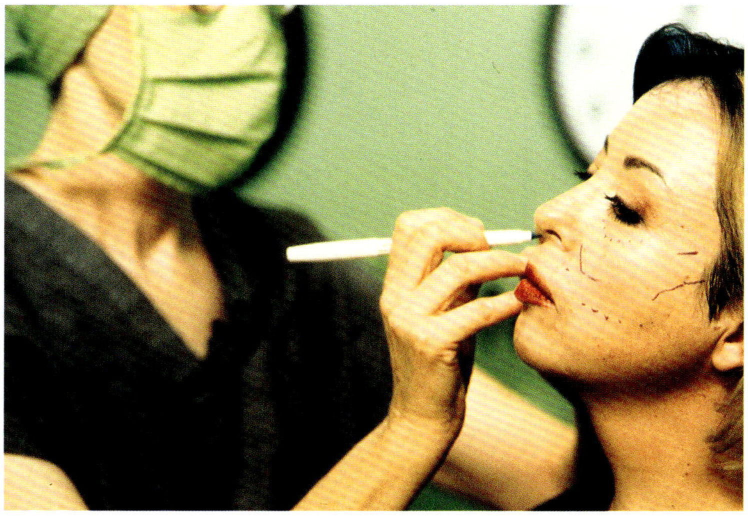

the medical profession, and our relationship to the ownership and design of our bodies. An integral part of her work is to give audiences the opportunity to challenge her provocative self-portrait and to question their own responses, by appearing in public debates and conferences.

Her actions and her willingness voluntarily to undertake risk will doubtless contribute in the long term to the demystification of surgical action, and to the mastery of the skin. Her aim is to make her body-object the focus of a redesign that will denounce it, change and transform it, without undergoing harm, mutilation or suffering. However, though she tries to make the work as painless as possible, using anesthesia and analgesics, Orlan admits that the surgical processes and injections are uncomfortable. She views this endurance as part of the challenge of the performance: she must go beyond her own physical limits, like an athlete breaking a pain barrier, to achieve an ultimate goal.

With the advent of new technologies in medicine there has been a definite shift in the dynamic between doctor and patient. Orlan's performances expose this new relationship and address technological and political transformations. Medicine, like other dominant ideologies in the Western World, has traditionally functioned on a pastoral basis: patients have looked to the medical profession for guidance and care over their ailing bodies and minds. Current society, having lost faith in the restorative powers of religion, has turned its face towards its doctors for reassurance that, even in the face of a technological era, we are still human. In light of these increasing pressures, doctors have responded by recruiting the assistance of new technology to assert the divinity of their science. Clinical diagnoses are justified through measurements and esoteric patterns generated by medical imaging systems. These provide a safeguard against human error. The 'body' has become digitised and stored on computers as a virtual corpse, where static images of humanity are awaiting reanimation and retranslation into living human flesh. The truth of the image is, however, imperfect since the anatomical designs originate not from a 'real' body but from a range of anatomical templates from different patients.

Orlan's understanding of carnality questions these templates and anatomical standards. Her enquiry warns that by religiously accepting the values of science, by our unquestioning sacrifice to medical institutions and our compliance in being dissected, ruptured, categorised and examined through medical microscopes, we are in danger of becoming sterile. And her confrontation does not stop at the borders of the flesh and body tissues: she believes that individuality resides even in the cell systems, fluids and genes that direct the molecular structures of the tissues. These units themselves, although invisible to the naked eye, being integral to the body architecture, are objects capable of redesign.

In order to demonstrate this, Orlan has preserved all the cells, tissues, blood and bodily trimmings from her six operations as reliquaries. An excerpt of a text by Michel Serres

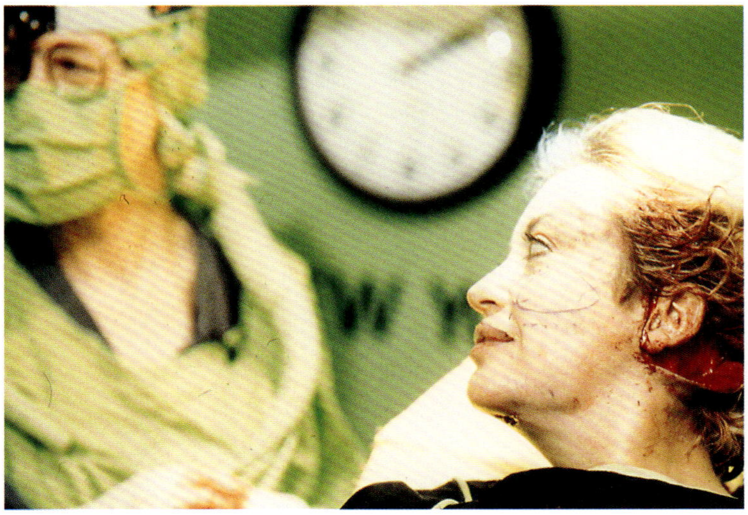

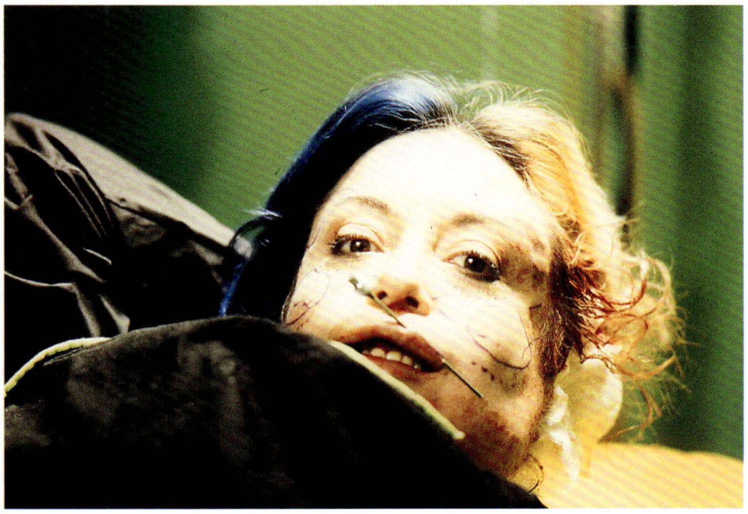

served as the framework for the operation/performance and is engraved on a series of bullet-proof glass tablets with a central receptacle containing 20 grams of her flesh:

> The current tattooed monster, ambidextrous, hermaphroditic and mulatto, what can it make us see now under its skin? Yes blood and flesh. Science speaks of organs, functions, of cells and molecules only to admit at last that it is high time we stopped speaking of life in laboratories; but science never mentions the flesh, which, quite rightly, signifies the conflation, here and now, in a specific site of the body, of muscles and blood, of skin and hair, bones, nerves and diverse functions, that inextricably binds that which pertinent knowledge analyses.

The aim of this operation/performance was to produce as many reliquaries as possible until no more flesh was available. In this act of self-depletion and eventual annihilation, Orlan paradoxically created potential sources of future reincarnation: each of the tissue samples contain her DNA providing an alternative, laboratory text for her own cloning and physical resurrection.The internal clock that induces ageing and metamorphosis, DNA, is also cited as an enemy: 'My work is blasphemous . . . I fight against God and DNA', she has said, challenging the notion that it is wrong to change the body that has been allocated to us by God or by genetic lottery, issued by chance and not design. The same society that objects to her interventions either stares at disfigurement or refuses to look at it at all, considering it a deviation from an ideological template of aesthetics to be 'mutant' or disfigured, and associating these stigmata with negative human traits. Orlan confronts these conditioned responses as unacceptable.

An ability to change appearance and function is essential for coping with an increasingly stressful environment. Orlan reminds us that we are able to transform the body, which in any case will inevitably change according to evolutionary laws. She turns this inevitability into a controlling force by deliberately making her body foreign. Her future design is selected from a catalogue of possibilities that respect science and anatomy but do not obey them implicitly. Using these guidelines she follows a new course of identity, breaking the skin and manipulating her natural morphology to excise redundant tissues and direct her own evolution. Following the transformation, the question will not be whether she is more or less beautiful, but whether the metamorphosis has been successful in matching her inner identity.

Orlan's Carnal Art is a celebration of new possibilities of being. Resisting the stereotypes of aesthetic tyranny she offers us inspirational, post-human body designs for new identities and appearances. *(Rachel Armstrong)*

Omnipresence, *1992*

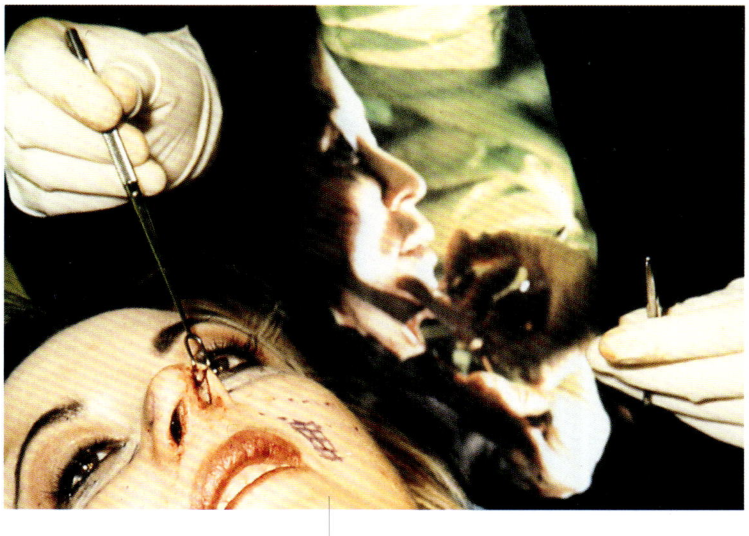

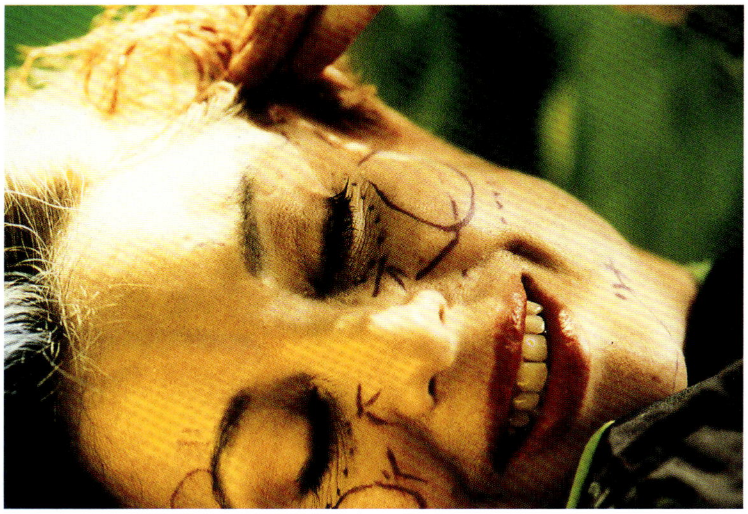

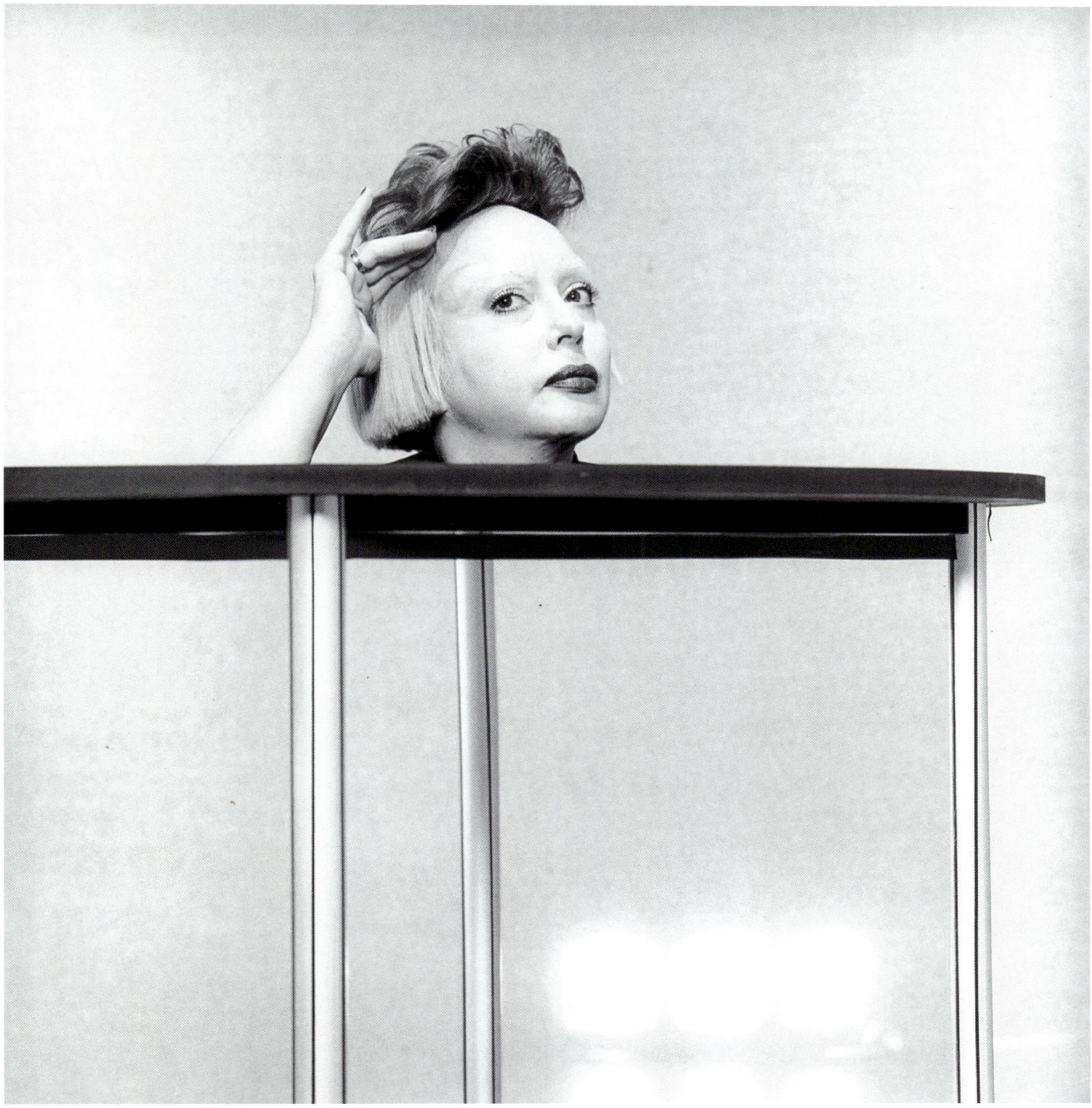

Woman with Head, *April 1996 (photo: Nicholas Sinclair)*

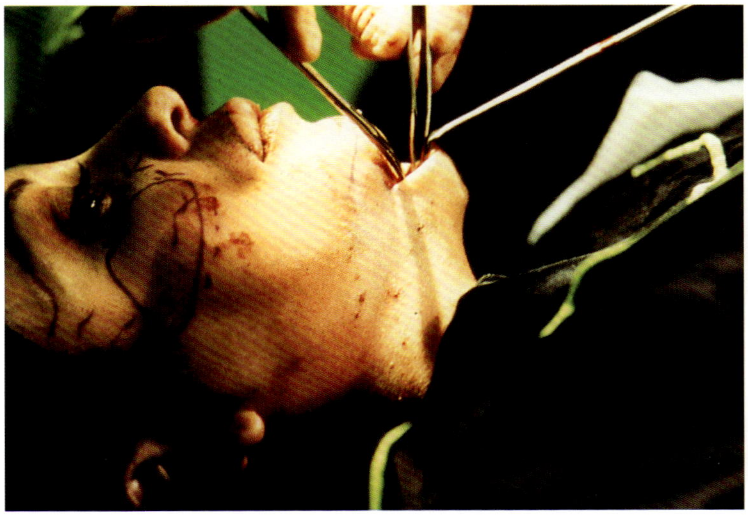

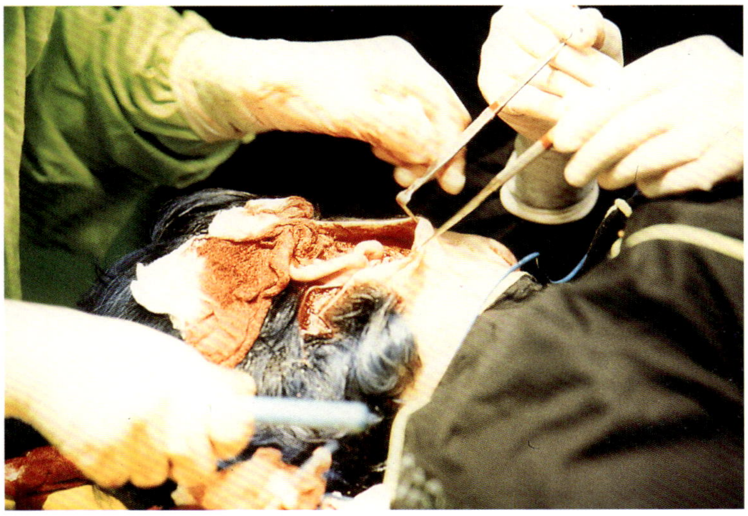

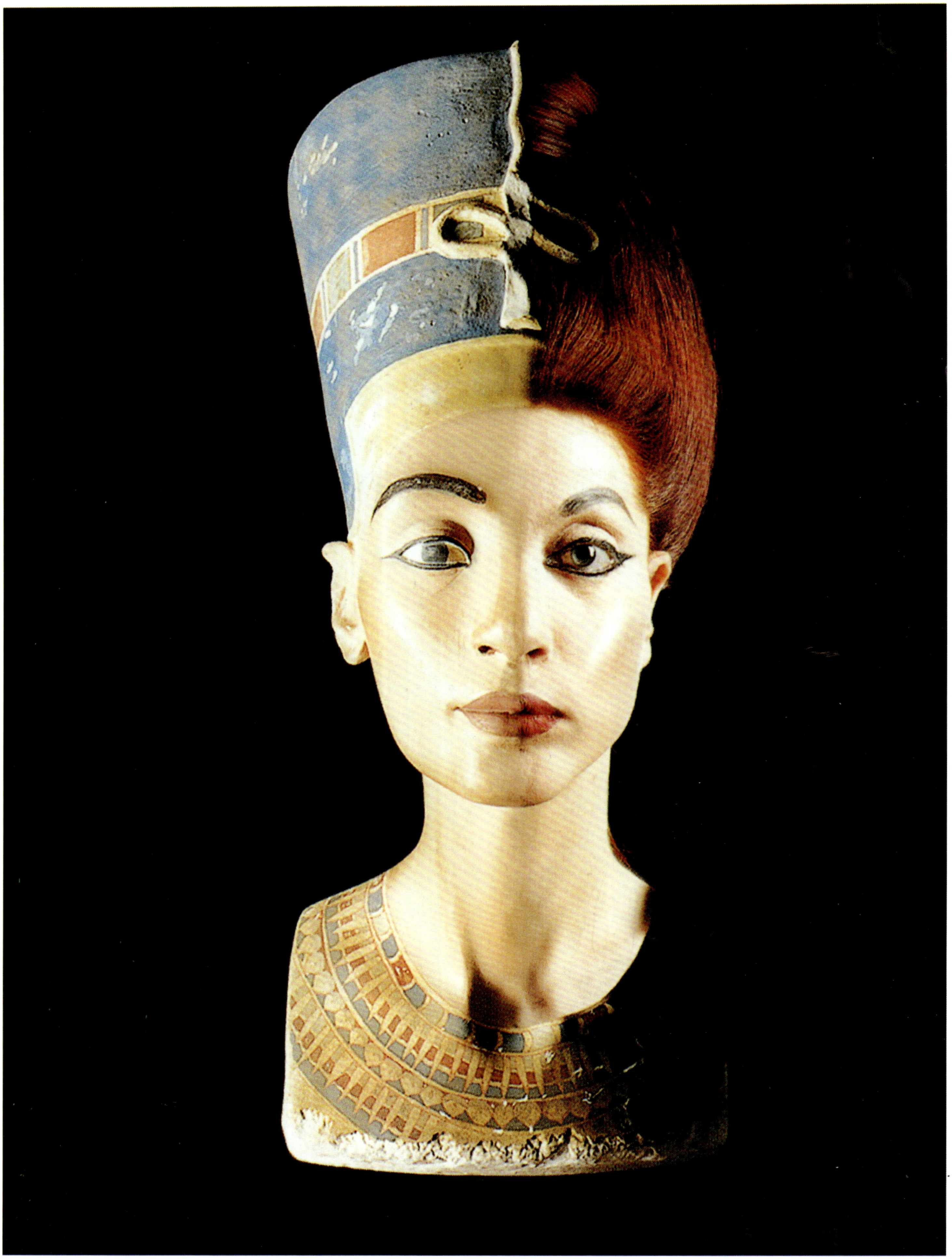

Elizabeth DYN (Photo: Jons Michael Voss)

ELIZABETH DYN

Soul Surgery

> *The importance of appearance is a curiously neglected aspect of human psychology. While the history of fashion is itself a testament to human insecurity, vanity and need, the psychologist has exhibited a greater fascination about what lies behind the visible and perhaps has missed an opportunity . . .* Professor Anthony Clare.

Elizabeth DYN is a mother, artist and writer who has committed her life to help others. During the past ten years she has shaped her natural face into a living sculpture, a project involving more than 20 operations.

Her goal is to complete by the end of the millennium an extraordinary, futuristic face that will give her greater inner strength and spirituality. Once it is finished she will begin a series of charitable works. She also aims to use The Face to create a new culture within the fashion industry by pioneering a form of modelling whose main objective is to enhance the clothes by expressing the beauty of the model's soul. Catwalk models would therefore be socially responsible, capable of expressing a new lifestyle through positive behaviour.

At the beginning of the project, she had no clear vision of how The Face would finally look, a situation leading to a large number of unnecessary operations that could have been avoided had she understood the surgical processes more clearly and been able to defend her intentions to the cosmetic surgeons.

Now her vision of The Face is that it should be without wrinkles, ethereal, delicate, almost angelic. This youthfulness will express her personality in a spiritual way, inspiring confidence in others that they too can remain young. Its overall geometry is based on an equilateral triangle running between the eyes and the mouth with a small, short and softly rounded chin. The eyes are strongly sliced upwards with a very short and concave nose. The mouth balances the other features with lips that are soft and feminine without the use of collagen.

Her search for The Face challenges the normal expectations of cosmetic surgery: the operations are more than a physical reconfiguring of the body, they act as a tool for self-development. Believing that beauty is part of the process of healing, Elizabeth has altruistic motives for spreading her message but acknowledges that not all women are at a suitable point in their own journeys of self discovery to use the same techniques. She hopes The Face will launch her as a businesswoman and an artist, enabling her videos and counselling service to spread her messages about the true meaning of beauty, nutritional health, tolerance, and the status of the ageing body.

Though she believes that we should resist the pressure put upon us by society to remain 'forever thirty', it is also her opinion that it is possible to keep the body and face in a state of perpetual youthfulness if we pay attention to inner harmony, nutrition, flexibility and natural bodily hygiene. Plastic surgery may be used as a delaying tactic to postpone the signs of ageing but without inner growth, the erasure of these stigmata is merely an assault on one's dignity.

Elizabeth models herself on the ancient Egyptian queen Nefertiti. Believing that there was only one Being – Aten – Nefertiti made it her mission to change the direction of a whole culture, moving the seat of high power, held by an old and very strong priesthood, to a new place, which led to alterations in many other cultural and social habits. Elizabeth dreamt that she was in the body of Nefertiti, able to see the events that were going on around her, but powerless to intervene because her soul and body had been separated. When she awoke she began to contemplate reincarnation and understood that the soul has a long past and future. (The pseudonym 'DYN' is a composite of the initials of her names in past lives). Reflecting on the parallels between the role of Nefertiti and her own ambitions, she decided on an approach that would use her creativity in combination with the tools of communication to show others how to evaluate and change their lives.

She believes that we are here on Earth to dissolve the restricting bonds that prevent us from developing our consciousness. The more we can help others to free themselves too, the more liberated our individual spirit becomes. If in any lifetime we neglect to sever the negative influences, we will have to continue to work on our development in another bodily form in the future. It is therefore an integral part of Elizabeth's development to be actively involved in projects that benefit others. Currently, she is helping women wishing to have cosmetic surgery to find doctors who understand their wishes.

Elizabeth sees herself as a pioneer of a new culture of people who are ready to take full responsibility for their choices and are adventurous enough to use modern techniques such as cosmetic surgery as a tool to express themselves. She believes that these techniques are literally able to stitch together the body and soul to create completely united individuals who are in touch with their inner and outer identities – a kind of surgery of the soul. *(Rachel Armstrong)*

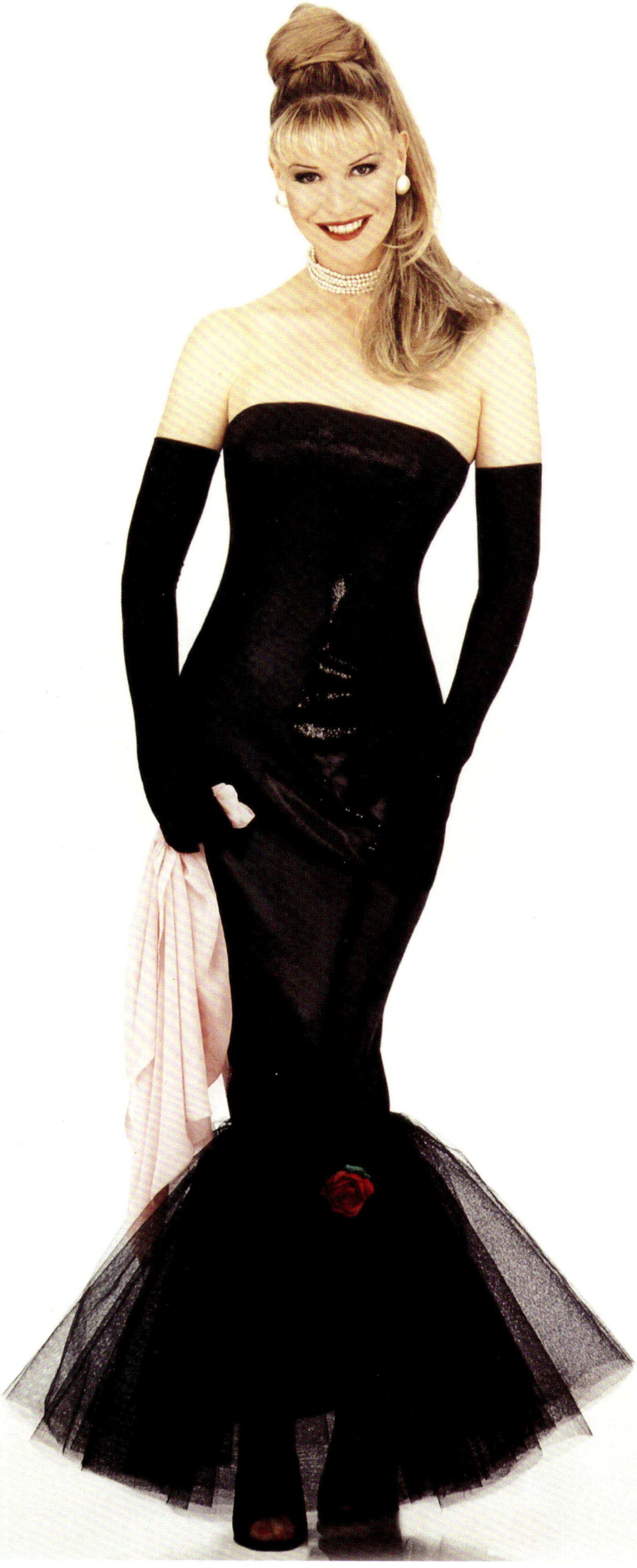

CINDY JACKSON

Beauty Is Power

Cindy Jackson is an artist and public personality who has successfully taken control of her own life by redesigning her body in the popular image of the Western feminine ideal. Having undergone more than 20 cosmetic procedures on her face and body to become as aesthetically pleasing as possible, she is now setting the standard of 'perfect beauty' for the new millennium.

For Cindy, beauty is harmony and symmetry. As a trained photographer, she has first-hand experience of techniques used to touch up the faces and bodies of supermodels in fashion magazine shoots. She believes that being 'naturally' beautiful is an irrational concept with various arbitrary meanings.

The idealised image she has constructed is an intentional masquerade for the purpose of gaining the power with which she was not born. She has come from rags to riches by presenting the right image at the right time. Beauty has empowered her to achieve desirability and admiration from those around her.

Many other women who seek this image lack the determination and vision to accomplish it, for fear of social retribution. Others, however, have copied her strategy, one even attempting to achieve an exact resemblance to Jackson herself, precipitating media discussion of the active recruitment of a 'Barbie Army'. Having found the power of beauty, Jackson does not mind being likened to a Barbie doll, seeing parallels in their stories. Born in 1952, as 'Lili', the creation of a cartoonist in Hamburg, Barbie also came from poor beginnings. Shortly afterwards she was made into a 30cm pornographic doll: designed for men, she was a symbol of illicit sex. Then in 1957, after purchasing three Lili dolls in Europe, Ruth Handler, the co-founder of Mattel, rechristened this icon of femininity, 'Barbie'. The new dolls, manufactured from 1959, represented purity and innocence. Now, Barbie is more than a doll to her fans, she is an inspirational symbol of the ideal woman. Interestingly, both Cindy and Barbie are popular amongst many women.

Cindy runs a global research and counselling service for women who want to know more about cosmetic surgery and who contact her for advice. She helps them choose surgeons, operations and the right image to give maximum appeal to their audience.

Throughout history, people have always striven to attain beauty through body modification. We are familiar with the 'primitive' practices of tattooing, branding, scarification and piercing and have accepted these as techniques to achieve beauty and power in certain societies. In contemporary life new technologies enable us to adorn our bodies in the pursuit of the same unnatural goals. The modifications afforded by modern surgery however, enable patients to appear genetically endowed with perfect features. According to Cindy Jackson, cosmetic surgery allows us to belong to society in a more integrated way than we have previously thought possible.

At this unique time, with technology, medicine and aesthetics interlinked through computers and nanobiology, Cindy believes that we are undergoing an evolutionary process. Surgery though, is not a panacea: it is still a somewhat primitive technique and has associated risks and post-operative complications.

Cindy sees cosmetic surgery as an alternative to cloning using non-genetic methods. Since the genetic codes of Dolly the sheep were identical to her 'Mother' in every way, and were not diluted by a male set of genes, they are closer to the attainment of immortality than the offspring of sexual partners. For this reason, Cindy has volunteered to be the first human to be cloned (her views on men and motherhood are as non-traditional as the process of cloning itself). Our struggle for the status of immortality is of great interest to Cindy, as are developments in neuroscience, artificial intelligence and body-preserving techniques such as cryonics. Beauty itself is part of the spectrum of ways in which, as individuals, we can achieve immortality.

The traditional method for preserving our own images of perfection is art. Trained as an artist, Jackson is currently working with various members of the art and academic community. Other artists are fascinated by her ability to wear her own self-portrait as a living mask and academics are still trying to find a category for her.

Cindy Jackson is a super beauty in an acquired position of power. She has physically manufactured her own dominant image in the recognition that it symbolises the command of influence and wealth. We all want to be part of the culture of beauty and Cindy is the first to show us how this can be realised in a practical way. *(Rachel Armstrong)*

For further inquiries write to :
Cindy Jackson, PO Box 3410, London N6 4EE

JAKE & DINOS MAKE MONSTERS

NICOLA TRISCOTT

Jake & Dinos Chapman are best known as the progenitors of a whole family of perniciously aberrant genetic monsters: unreasonable objects of reasonable parts. Their current creatures are sculpted dummies randomly composed of arms, legs, heads and genitals. The anatomical units are reworked and combined to form phenotypic mutations resulting in a series of works that have provoked debates around the subjects of neoteny, pornography and sterility.

These sculptures seem intent on exploiting what Douglas Fogle describes as 'the troubled zone of scientific uncertainty', in which lie strange biogenetic mutations, upsetting science's quest for rational order. 'Science is the system of homogeneity *par excellence* . . . the entire project of the modern scientific endeavour has been devoted to an all-out attempt to banish uncertainty' he says in the catalogue to the recent exhibition, 'Chapmanworld' at the ICA, London.

In future works the Chapmans intend to become true Dr Frankensteins, conducting mutational experiments on living or life-simulating systems.

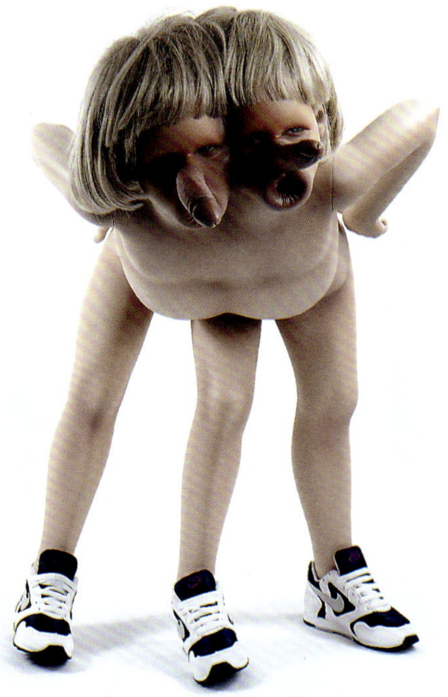

FuckFace Twin, *1995, fibreglass, resin, paint, 85 x 54 x 67cm*

FUCKFACE

JAKE CHAPMAN

When murder is not the soul provenance of adult misanthropy, and when misanthropy runs in your genes.

Dirty fucking outrageous Bulgeristic events focalise the community in the religious iconicity of violent spectacle: made religious by the fact that a child is not attributed with the same vertiginous hatred of others that bad adults are potential of: AND BECAUSE MINORS ARE NOT MATURE ENOUGH TO BE PATHOLOGICALLY IMMATURE LIKE SOME BAD ADULTS, THESE CHILDREN ARE THE EPITOME OF EVIL – ANOMALIES THAT EXIST IN A SOCIETY WHERE RELIGIOUS BELIEF HAS BEEN EXCHANGED FOR THE PERSUASIONS OF PATHOLOGY AND THE CLINIC. IF THEY EXHIBIT EVIL it is because the psychiatric police have failed miserably and the devil has returned digitally integrated within the socio-neural circuitry, stronger and more compulsive than ever . . . Why is a child presumed not to have murderous intent, or at least morbid experimental curiosity? (Read some graffiti written in red, non-toxic crayon on a wall which said 'Kill your parents, kill your teachers, kill yourself' – don't know why I mention this – thought you should know anyway – especially on the off-chance it might have a subliminal effect – you never KNOW.)

Fuckface

Once upon a time, right now, in a vast country far away lives a little person called Fuckface. Fuckface is the hatechild of universal parental icons Mummy and Daddy 'O'. But Fuckface is neither boy and neither girl, NO. Fuckface is just plain Fuckface. By birthright Fuckface is a disavowal organ, a loosely-spotted cockanalcunt. And given the wealthy plethora of genitalia proudly sprouting across the perfect anatomy of Fuckface's Mummy (who doesn't appear strictly mumsie), castration was never a concern for the helpless-infant-Fuckface, because Mummy's countertransference was never deficient. On the contrary, she was supplied with every luscious desiring apparatus, and, of course, Fuckface was never in competition with Daddy for Mummy's love because Fuckface's Daddy (who doesn't strike one as being particularly dadsie) isn't exactly phallic – with all his viral sphincters mapped across his pitted body. Consequently no hypertrophied SUPEREGO was ever formed in Fuckface's little big head.

'Who's been fucking **my** sphincters?' said Daddy.
'Who's been fucking **my** cocks and cunts?' said Mummy.
'Who's been fucking **my** face?' said Fuckface in a muffled voice.
'Why, how silly, **we all have**!' they laughed.

Social status is defined by crime: either you are an entrepreneurial criminal who is merely interested in the expansion of your own immediate (local) economic status, or you are a state criminal who abides by the law of judicial consensus yet sanctions every economic, techno-hegemonic, financial and military crime enacted on your behalf over geopolitically weaker sites. Superman rules.

Rule (1): our pleasure becomes de-motivated by the threat of entropy = energumen experienced through labour time: industrial bodies are obsolete, and so are the technologies of our industrialised synaptic stimulation that propelled them into former servitude: boredom and inertia are LIVE (low ebb) options procured on behalf of advanced leisure time and narcosis: WE tentatively predict mass psychosis in the future; moral imperatives latent in historical consciousness will blur in the civic hallucinatory state; the polity will slip into an oceanic equilibrium, as privileged narco-aesthetic-culture experimentation (formerly privileged to mind-expanding, trust-fund hippies and Byronic aesthetes) will become de-marginalised away from such culture edifying psycho-nauts, elevating the frictions of everyday grind to merge with dream time of reified catatonic States caused through conscious-prolapse (thought haemorrhage). Spoilt existential nausea and the genesis myth of SINGULARITY itself (the unit myth of organic autonomy) will dissolve LIKE disprin, annihilating headaches forever. The subject will achieve antique redundancy on both biological and metaphoric levels, splitting in shreds to infinity. In fact, metaphor collapses first because its axiomatic hold over the subject in producing models of performativity by which the subject traverses/feels-out the field of essential substances and properties of social life (family, gender, domestic sickness, credit finance) will no longer be properly exchangeable since reality testing will subside into a poly-magnate of ripe biogenetic shit synthetically replicated to make the remaining post-obsolescent human organisms feel at home in the mire.

Cogito
errrrrrrrrrrrrrrrrrrrrrrrrrrrrrrrrrrrrrrrrrrrrrrrrrrr
go
sum.

Nuclear subjectivity? The demand for work? (Delegating the pressures of compelling physical laws that weigh speculation to gravity? The qualitative demand for work ((energy)), where

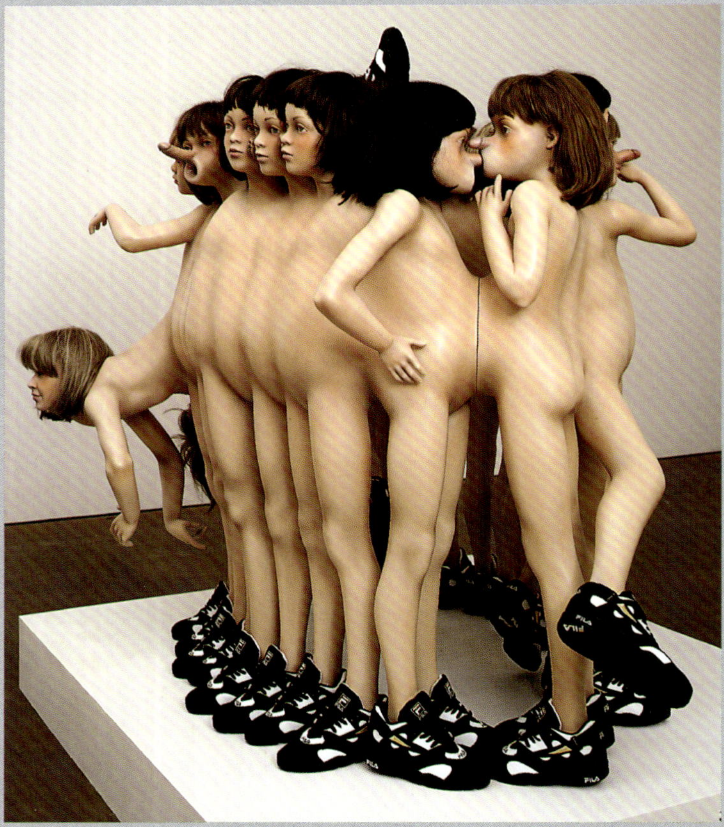

BEING is defined by DOING? The world of industries necessitatED being to be defined by doing, the imperative to measure subjectivity by productivity.) 'What do you Do?'. 'Nothing (=death)'.

This imperative has been jilted by the chaotic calibrations of adult-infantile-professional-play, which burns accumulated resources without reserve (playing the stock market can hardly be judged as being consistent with work). The familial-social model has subsided ineradicably into alienation ('let's crash the creche'), metamorphosing from the social-industrial productive model (little Janet and John hydraulic units) to cast an ambitious pathological silhouette, a nemesis of active narcolepsy, productive of random thoughts driven by detached desire-fragments: a post-conscious quantum phenomenon, broadcasting radioactive particle decay like BO-corporeal odour, subject-stench, Necrophobia; Newtonian-neurosis = the fear of **entropy** (=death).

Undeniably, the following quotation is brimming with an overtly human emancipatory wish, but beyond Lyotard's limited desire to '**rid ourselves of the will to be cured**' its physical momentum, itself a Newtonian ideal of frictionless motion gliding between the inhibiting and productive sources of the organism, implies a non-resistant depletion of psychic resistances, which in its neuro-vehicular movement supposes the dissolution of the resistances-building-up-in-neurosis combining to secrete 'the self' within the terms of reference that is its sovereign domicile: N-E-U-R-O-SSSSSSSSSSSSSSSS-I-S. or as Deleuze so aptly screamed as he flew from his hospital window:

'all forces are dominant or dominated' splat

God's terminal drop zone

God turns a corner and falls into a dirty great black hole. From the unmistakable stench he recognises the cavernous interior of Sade's arsehole. And now, Sade is squatting to shit, capillaries tingling his cheeks from the colossal strain, and the Omnipotent is expelled to Earth to steam stunned and motionless in the ground frost. Pulling his trousers up Sade runs off, the hysterical laugh diminishing into the distance, eager to set another trap for his buffoon.

For the benefit of demonstrating the sensual pleasure in nihilistic science, Human(e)kind has endeavoured industrially to dilate God's anus, to ritually incite a flamboyant unholy enema, a religious purge, manufacturing the necessary toxins, the chemical incense, to force the Ozone layer to relax its tight clench upon the Earth. In face of this blatantly Sadistic act, the neurosis of ecology anxiety is a tacit eroticising of the death drive that similarly makes the mass-murderer drag his heels and cry like a baby as he is taken to be exterminated. Boo, hoo, hoo.

Zygotic acceleraton, de-sublimated libidinal model (enlarged 1,000), *1995, fibreglass, resin, paint*

DANDY DUST

A HANS SCHEIRL

Dandy Dust is my feature-length, avant-garde science-fiction film. The setting is a cybernetic planetary system, 3075, whose inhabitants are comic-strip characters inspired by splatter cinema, Hong Kong gangster movies, Japanese live-action films and Manga.

The story involves the adventures of a dysfunctional family and the search for self-determination of the fluid-gendered hero, Dandy. This journey of self-discovery is depicted in a series of adventures full of dangers and strange creatures, bizarre monsters and mutants.

Main Characters and Time/Spaces in order of appearance:

Planet of White Dust
The organism on which the 3W (White Western War) cultures are fighting its historic battles, made of bones and bone-flour. In its centre, a heavy-metal ammunition machine, driven by the pelvis-force of The Workers, is ejaculating bullets.

The Mothership
A flying graveyard – Super-Mother Cyniborg's after-life space ship. Inside the old family tombs, the uterus of the ship, Cyniborg has set up a sick-bay tent where she is busy stitching back together deceased ancestors to restore the family tree.

The Twins: Mao and Lisa
Dandy Dust's sisters. Never allowed to be born they are working for Cyniborg as her body-part collectors. They operate with precision, silently and dispassionately.

Dust
Fluid-gendered time traveller with a memory problem.

3075
Cybernetic City/Planet/Creature. Its humanoid inhabitants exchange bodily fluids with the city and each other inside the bladder of the planet's body. They communicate via chemical reactions and electronic mini-rhythms or sexual trembling. The planet is self-regenerative; its time pattern is the year 3075 AD.

Sir Sidore
Dandy's father, English Landowner. His time/space is the Planet of Blood & Swelling. He always caries a huge hard-on in his breeches. He is repressed, weak and bitchy.

Cyniborg XVII – Duchess of Loft & Spire
Dandy's mother. A descendant of an old aristocratic family, she is obsessed with genealogical power. She hates her husband Sir Sidore for his lack of masculinity and repressed randiness. She wears 16-inch heels and a bishop's mitre. She has pointed claw fingers and one very long arm.

The Mummy
A shape-shifted, after-life persona of Dandy whose goal is to unite with Dust to get his/her body back. Traumatic experiences and death have made the Mummy cynical and sadistic.

The Flame
A shape-shifted persona of the Mummy.

Theodora
A supportive great aunt of Dandy. A witty, Mesopotamian, mathematical genius, restored and rejuvenated by Cyniborg.

Dandy
Dust's past self. Lives on father Sir Sidore's Planet of Blood &

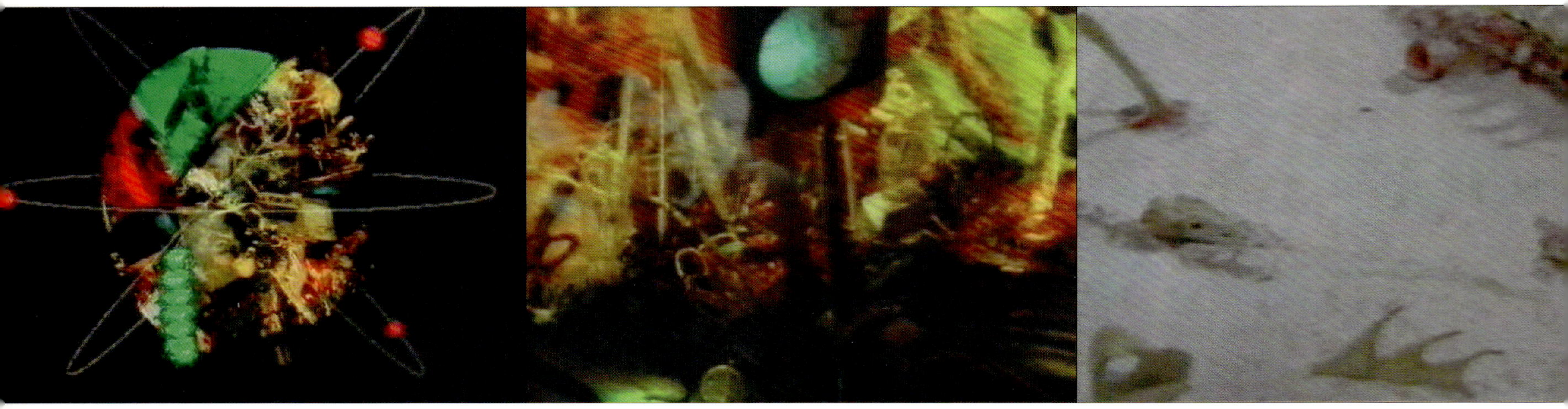

Swelling. The mysterious death of mother Cyniborg makes him/her a lonely, melancholic adolescent. Enjoys hunting and collecting small dead animals.

Planet of Blood & Swelling
Aristocratic English time/space. The forests and the skies are the intricate bloodvessel landscape inside Sir Sidore's anus.

Spider
A creature with special powers of time transportation and hypnosis. Shape shifts from a tiny, hairy ball to a humanoid with a big vagina on her belly. Likes to suck genital juices.

Synopsis
Cruel wars are raging through the centuries on the Planet of White Dust. Super-mother Cyniborg lands with The Mothership onto the corpse-strewn planet: she is searching for missing relatives. The Twins are at her side, doing the dirty work.

We hear screams. A soldier is raping another soldier, holding a gun in his mouth. Zack! An arrow hits the rapist's body. Zack! Another arrow pierces his trigger finger, Out of all his body openings sperm gushes in abundance. Finally all that is left of him is a white puddle that gets soaked up into the bone-sand and dries instantly. A gust of wind carries it up into the sky. The dust trails into patterns and forms the title 'Dandy Dust'.

Dust comes flying out of the starry sky. Noticing planet 3075 below, s/he belly-flops down. A coil whirls itself around her/him and s/he is catapulted into the supervision tower of 3075 where two people in white overalls and dust-masks open her/his skull to steal her/his memory unit.

Dust stumbles across 3075's landscape. A cyber-surgeon takes Dust's suit in exchange for re-wiring the loose cables in her/his head.

The city/planet 3075 is introduced by the narrator. Dust becomes part of its cybernetic structure.

While plugged into the bladder of the city, Dust has an apparition: s/he sees Dandy, his/her past self in 18th-century clothes. Dust creates a fetishistic, cyber-fantasy-space art-class that can be visited through the use of a special drug. It is an end-of-19th-century, gentlemen-only art academy. The students draw nudes and cruise each other. Sir Sidore enters wearing a bloody mask. Dust flirts with him. Out of a drawing another strange figure appears: Cyniborg! She is accompanied by her two daughters, the Twins, Mao and Lisa.

The Warden approaches them. Cyniborg is not wearing the right dress code since no high heels are permitted in class. She retaliates. In the scuffle, the newcomers start to bleed from old wounds, and blood splatters everywhere: they are zombies. The incident, which takes place in virtual reality fantasy, leaves Dust confused.

A winged coffin forces itself through the urinary system of 3075 into the bladder. A crumbling, dusty figure steps out: the Mummy. S/he claims to be Dust's better half and informs Dust that the bleeding zombies are their common parents, committed to returning them to the vaults of the family crypt. The Mummy argues that only their unification would bring back lost memories and the strength to fight the crazed family.

This sounds to Dust like old fashioned love-crap. In a violent argument, Dust destroys the Mummy who turns out to be only a hollow shell of bandages. Dust tramples on these in anger and throws them onto a pile of rusty metal.

A tiny flame is all that is left of the Mummy.

Dust is killed by the Twins and brought to Super-Mother Cyniborg's space-ship, the Mothership.

Cyniborg puts Dust into heavy shackles. She is after his/her eggs: the fancy family jewels.

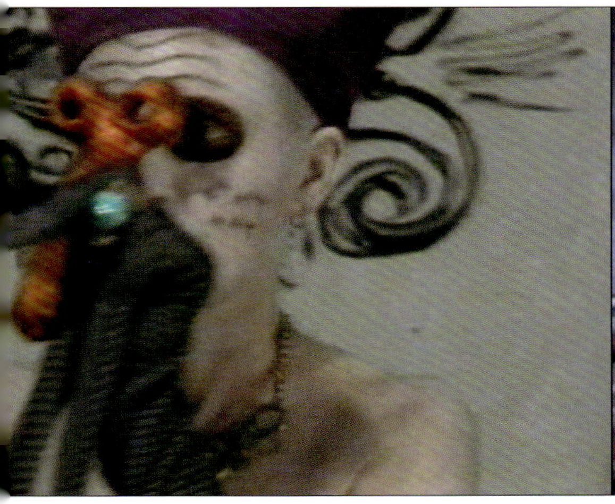

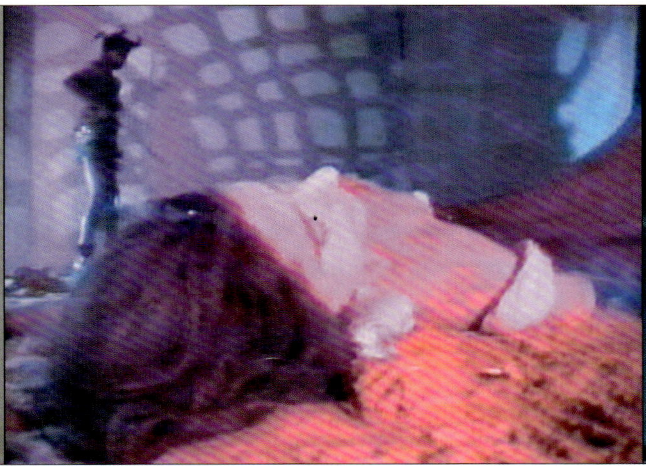

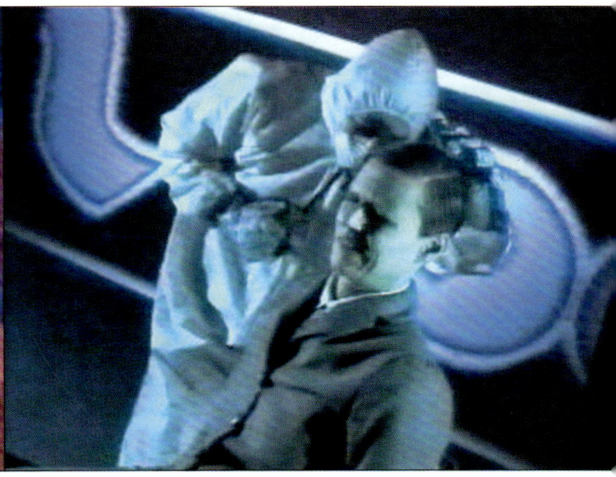

The Flame, who is floating in the air above the neck-hole of a hollow suit, helps Dust to get free and introduces her/him to old aunt Theodora, a newly stitched up ancestor.

Flame and Dust watch their lives in flashback on Theodora's orange monitor.

1779, a cold day in an English country cemetery. Dandy mourns her/his mother's mysterious early death.

Dandy, always an androgynous child, has been raised by father Sir Sidore as a boy. A repressed sexual attraction underlies their righteous manners.

On one of her/his walks through the romantic bloodvessel countryside, Dandy carves a hole in her/his favourite book on time-travel in a last effort to find a way out of this depressing time/space. As the night progresses with Gothic spookiness, a hairy ball hatches out of the book: Spider.

Spider lures Dandy into a deal: in exchange for sexual favours s/he promises to take Dandy on a trip into the future to visit sexually fluid future-self Dust. Dandy agrees and appears in the bladder of the planet of 3075 for a short moment to see Dust in electro-chemical ecstasy.

This incident clicks Dandy out of her/his repression into horny megalomaniac madness. At a dinner party given by Sir Sidore, s/he stuffs food into her/his pants, bites a lady (Lady-D) on her leg, makes a pass at her/his father, pulls the tablecloth from the table so that the wine and food hits the noble guests, catches a mouse and eats it. Dandy pursues the outraged, Lady-D and forces a bloody kiss on her mouth. Lady-D severely punishes Dandy in front of everyone and passionately kisses her/him. Sir Sidore's desire and hate are unleashed. Sir Sidore rapes Dandy in the stables with a drill-dildo. Dandy shoots her/his father.

The Twins transform the remains of Dandy and Sir Sidore into TV-particles to transport them through space to the Mothership. During their journey the Twins force Dandy (or rather her/his electronic identity) to remember the horrible family tragedy:

The child Dandy is loved and spoiled by both her/his parents. They themselves hate each other profoundly and make the child into their mediator.

When Dandy the child hears the Mother Cyniborg is pregnant with twins, s/he kills Cyniborg. Instant guilt erases the murder from the child's memory and divides her/him into two: one of them lifts off into the air to become Dust the time-traveller, the other one stays to become the adolescent Dandy.

At Mother Cyniborg's funeral, the earth trembles: enraged by her murder, Cyniborg pulls out of the ground.

Terribly moved by the traumatic events of their common childhood, Dust and Flame unite and become Dandy Dust. Now exuberant, s/he dances around and weeps in desperation and joy. Theodora, who has been busy unscrewing her room to get away from Cyniborg's claws and the morbid stuffiness of the ship, comes down the ladder.

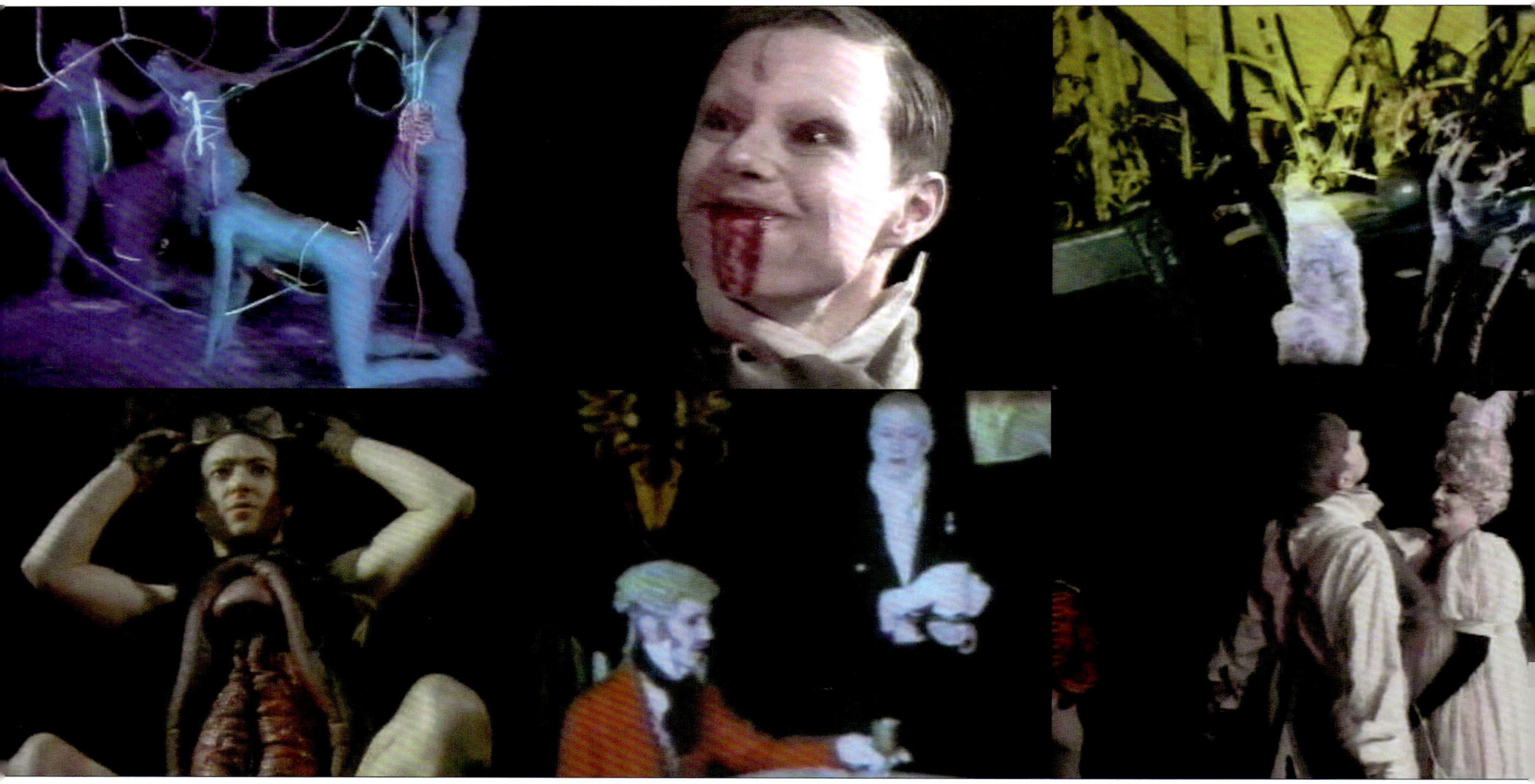

Theodora notices Dandy Dust and asks who s/he is. Dandy Dust, nearly speechless, sobs out her/his name. Theodora takes her/him into her arms and fetches a drink. Soon Dandy Dust is ready to watch the rest of the flashback:

The Twins land their caterpillar on the drive-way of the Mothership with Dandy and Sir Sidore's TV-particles in their scoops.

After manipulation of the electronic remains of her loved ones, Cyniborg transforms them back into their heavily abused bodies. Dandy gets a new vagina.

Bandaged from head to toe, Dandy becomes the Mummy and escapes the Mothership with the help of Theodora who gives her/him a flying coffin. The flashback ends.

Now Dandy Dust is ready to face the tyrannical mother and finally say goodbye to zombie-family life. S/he walks through the vaults of the crypt repeating her/his mantra:

'I am Dandy Dust. From now on I will take my life into my own hands. I love you all dearly but I have to go my own path. Believe me . . .'

Out of the shadows of the crypt, hands are grabbing her/him; the family members pick Dandy Dust up and strap her/him into Cyniborg's gynaecological chair. They approach with big, fat, ejaculating dildos on sticks.

Cyniborg proclaims that she will put and end to the sloppy family patchwork and vows to breed clean and intelligent offspring to preserve the purity of the family future.

Dandy Dust retorts that s/he cannot be reached and is without fear. S/he chides them all and invites her/his family to feed on her/his rivers to see for themselves that s/he is burning.

Suddenly the family members step back; one after the other they tear off their masks: with horror, Dandy Dust realises that they are all her/his self.

The Mothership trembles. The camera approaches a red inflatable tent on the bow of the Mothership. The Twin Lisa is sitting inside, frantically touching the walls of the tent, which are covered with chemical formulas.

At the same time, in the supervision tower of 3075, Twin sister Mao is supervising the last seconds of the count-down to the planetary quantum leap, which coincides with their successful transfer to the planet 3075 that was grown parthenogenetically from their own eggs.

Theodora, who has just managed to unscrew her room from the Mothership, pulls down again and shouts at Mao to stop playing with the old girl's blood pressure.

Mao pulls Theodora into the tent and tells her that they are going for a ride. The walls are flapping madly. The Mothership comes and flushes everyone out into black space. The characters wave at each other once more before they fly off towards new adventures.

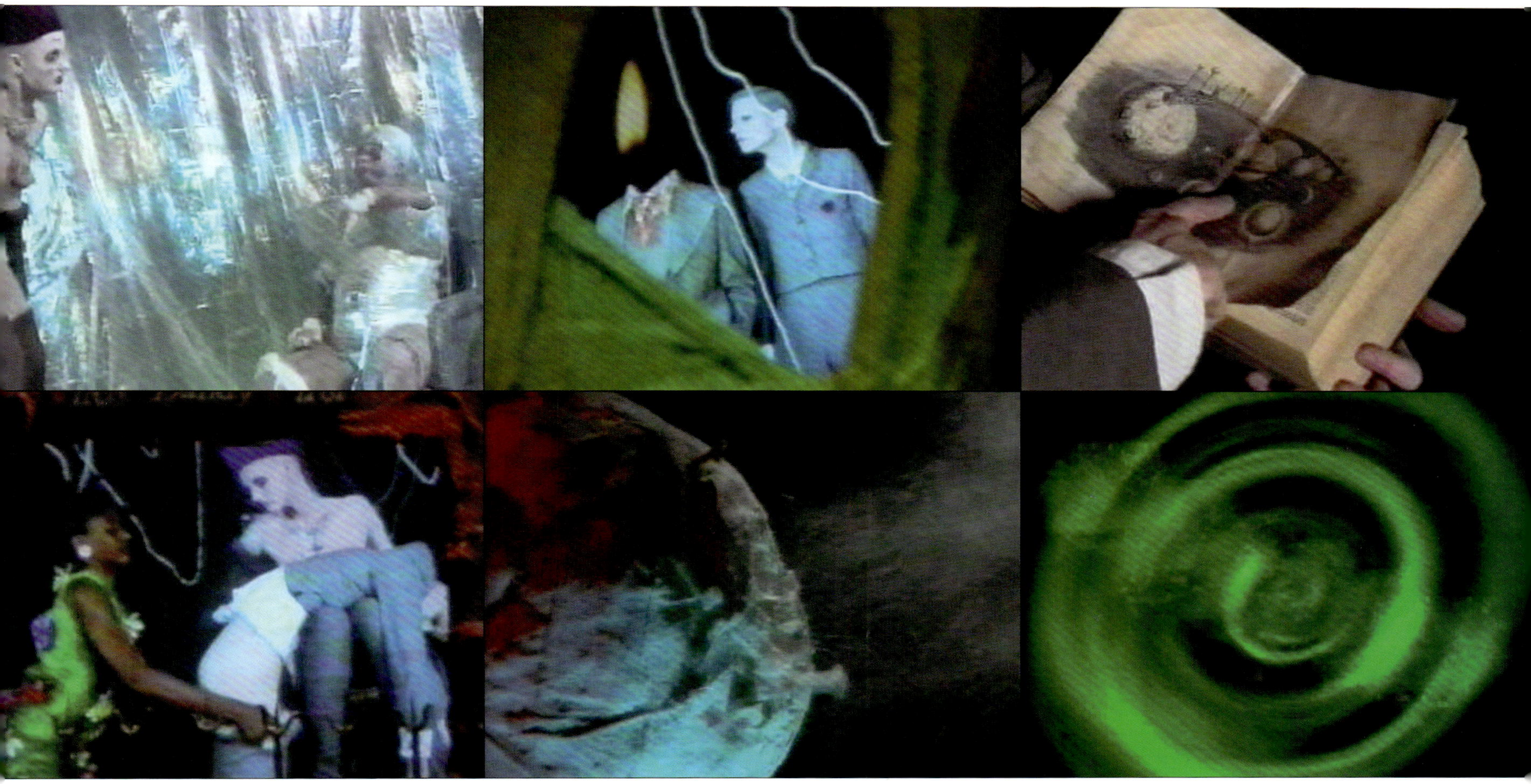

BITCH MUTANT MANIFESTO

VNS MATRIX

The atomic wind catches your wings and you are propelled backwards into
the future, an entity time travelling through the late 20th century, a space
case,
an alien angel maybe, looking down the deep throat of a million
catastrophes.

screenflash of a millionmillion conscious machines

burns brilliant

users caught in the static blitz of carrier fire

unseeing the download that scribbles on their burntout retinas

seize in postreal epileptic bliss

eat code and die

Sucked in, down through a vortex of banality. You have just missed the
20th century. You are on the brink of the millennium – which one –
what does it matter?

It's the cross dissolve that's captivating. The hot contagion of
millennia fever fuses retro with futro, catapulting bodies with organs
into technotopia . . .
where code dictates pleasure and satisfies desire.

Pretty pretty applets adorn my throat. I am strings of binary. I am
pure artifice.
Read only my memories. Upload me into your pornographic imagination.
Write me.

Identity explodes in multiple morphings and infiltrates the system at
root.

Unnameable parts of no whole short circuit the code recognition
programs
flipping surveillance agents into hyperdrive which spew out millions of
bits of
corrupt data as they seize in fits of schizophrenic panic and trip on
terror.

So what's the new millennium got to offer the dirty modemless masses?
Ubiquitous fresh water? Simulation has its limits. Are the artists of
oppressed nations on a parallel agenda? Perhaps it is just natural
selection?

The net's the parthenogenetic bitch-mutant feral child of big daddy
mainframe. She's out of of control, Kevin, she's the sociopathic
emergent system.
Lock up your children, gaffer tape the cunt's mouth and shove a rat up
her arse.

We're <>verging on the insane and the vandals are swarming.
Extend my phenotype, baby, give me some of that hot black javamagic
you're always bragging about. (I straddle my modem.) The extropians were
wrong,
there's some things you can't transcend.

The pleasure's in the dematerialisation. The devolution of desire.

We are the malignant accident which fell into your system while you were
sleeping. And when you wake we will terminate your digital delusions,
hijacking your impeccable software.

Your fingers probe my neural network. The tingling sensation in the tips
of your fingers are my synapses responding to your touch. It's not chemistry,
it's electric. Stop fingering me.

Don't ever stop fingering my suppurating holes, extending my boundary
but in cipherspace there are no bounds
BUT IN SPIRALSPACE THERE IS NO THEY
there is only *us*

Trying to flee the binary I enter the chromozone which is not one
XXYXXYXXYXXYXXYXXYXXYXXYXXYXXYXXYXXYXXYXX
genderfuck me baby
resistance is futile

entice me splice me map my ABANDONED genome as your project
artificially involve me
i wanna live forever
upload me in yr shiny shiny PVC future

SUCK MY CODE

Subject X says transcendence lies at the limit of worlds, where now and now,
here and elsewhere, text and membrane impact.

Where truth evaporates Where nothing is certain There are no maps

The limit is NO CARRIER, the sudden shock of no contact, reaching out to
touch but the skin is cold . . .

The limit is permission denied, vision doubled, and flesh necrotic.

Where truth evaporates Where nothing is certain There are no maps

The limit is NO CARRIER, the sudden shock of no contact, reaching out to
touch but the skin is cold . . .

The limit is permission denied, vision doubled, and flesh necrotic.

Command line error

Heavy eyelids fold over my pupils, like curtains of lead. Hot ice kisses
my synapses with an (ec)static rush. My system is nervous, neurons
screaming – spiralling towards the singularity. Floating in ether, my body implodes.

I become the FIRE.

Flame me if you dare.

PETER GABRIEL WITH REAL WORLD

Eve

Under the guidance of Peter Gabriel, the Real World multimedia team have produced a parallel world in the game *Eve*. The player is displaced into a fourth dimension in the vastness of cyberspace, linked to their immediate reality through the computer interface.

Just as in science-fiction stories, the protagonists visit parallel earths with alternative histories, the adventurer of *Eve* is translocated into a fantasy land set in the information-rich environment of the digital domain. The journey allows the player (or hero) to find the self-fulfilment that evades them in everyday life through their interaction with the challenges and encounters of the game. Many of these experiences are carefully designed to provoke players to question their assumptions of the natural laws and scientific principles on which our understanding of the world is based and perhaps imagine alternative human evolutionary pathways.

Gabriel is already renowned for his unique use of technology in his music videos, his previous title *Explora*, and his quest to bring together a diverse number of cultures through technology in the pioneering World of Music And Dance (WOMAD). Here, he invites the player to rethink perspectives on human relationships and to imagine and build a new world from scratch. His team are not scientists, but radical thinkers, artists and musicians.

'When an artist touches something real they have created a tool. Multimedia is a wonderful place for investigating art as a tool-kit . . . I'm very keen in seeing my work deconstructed', Gabriel has said. His parallel world entices and surprises the adventurer with its own internal logic whilst maintaining a rich, organic feel. The starting-pont for *Eve* was accepting that, in this life, Paradise is lost to us. The focus of the experience is to re-examine the relationship between man and woman through the gender archetypes Adam and Eve.

Reunification of the lovers after their separation and expulsion from Eden is only possible after the player has navigated four parallel worlds designed by the artists Helen Chadwick, Yayoi Kusama, Cathy de Monchaux, and Nils-Udo. Each offers the player a magical environment that interweaves contemporary art and music, containing fragments based on previously unreleased versions of Peter Gabriel's songs 'Come Talk to Me', 'Shaking the Tree', 'In Your Eyes' and 'Passion'.

As the players progress through *Eve*, they are able to collect Gabriel's musical samples (represented by an image from one of the artists' works), which they can ultimately use to mix into musical tracks. They will also discover a series of interviews in which people discuss the characteristics they find interesting and attractive in others. Each time adventurers visit the world, they encounter different experiences, which they can represent on their personalised sound-tracks.

The computer environment is a powerful medium for exploration, not only providing a new dimension for adventure but also offering a unique inbuilt intelligence that takes the viewer away from the mundane and into a radically different world. This world of new juxtapositions and different audio-visual associations reflects the way in which Gabriel predicts our current communications technologies will evolve:

> I believe that in the next few years kids will develop their own multimedia language – free of the barriers of text and translation – that we will ultimately adopt. All around the world video, music, and graphic images are beginning to be pumped down phone lines and the internet . . . relationships, spiritual matter and nature will all become important new focal points for technology and perhaps that will allow us to relate better to others.

(Rachel Armstrong)

RIGHT: Nils-Udo, Waternest (Eve)*; OVERLEAF: More images from* Eve*, by Peter Gabriel with Real World Productions in collaboration with the artists Helen Chadwick, Yayoi Kusama, Cathy de Monchaux and Nils-Udo*

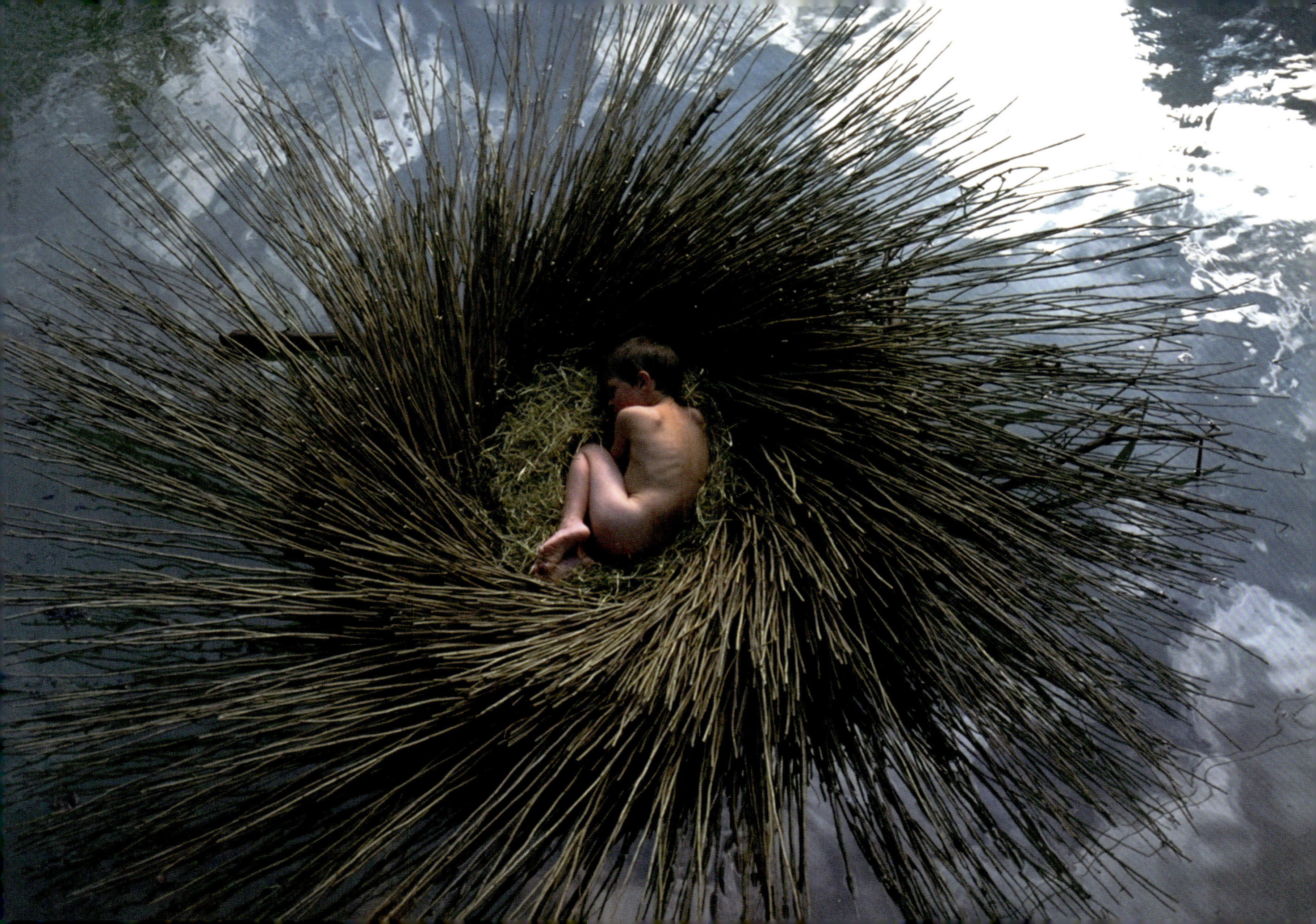

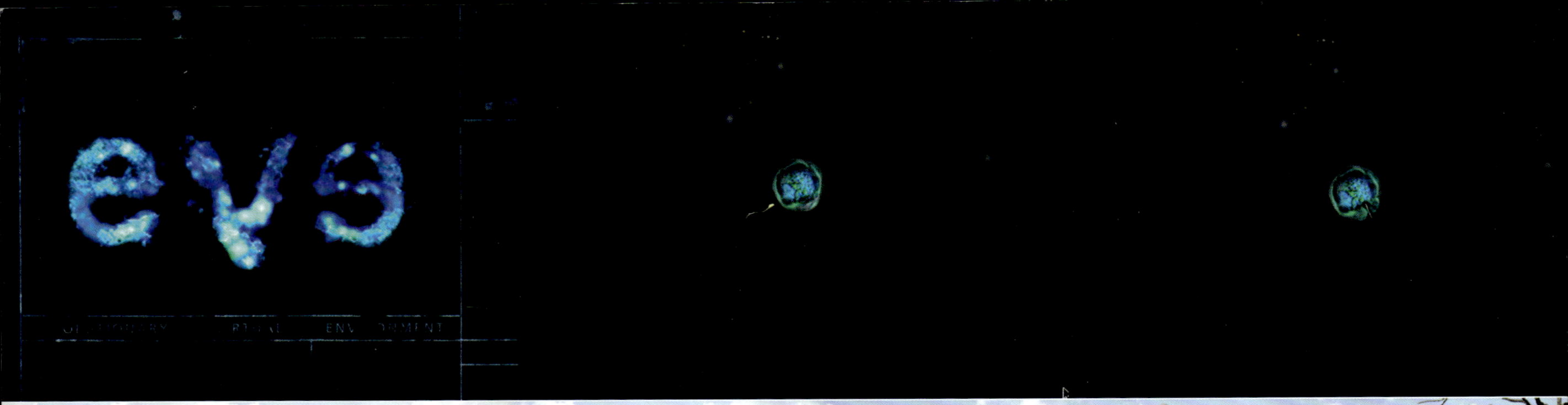
eve

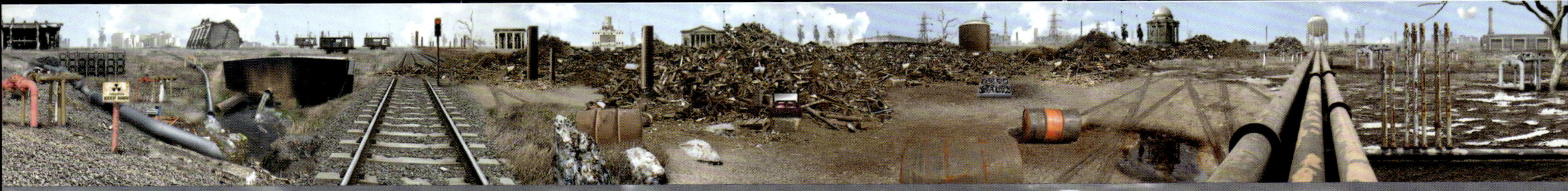

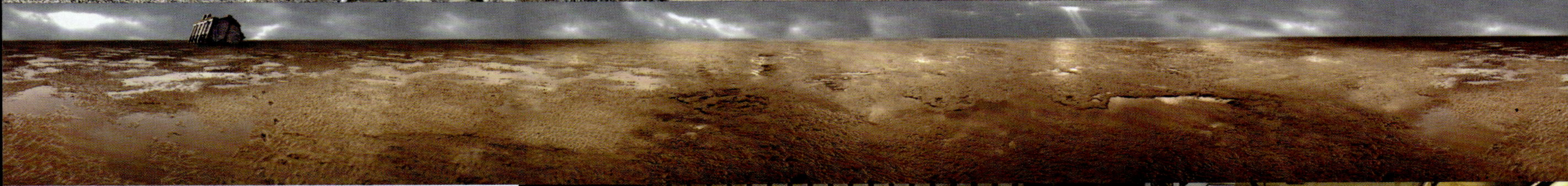

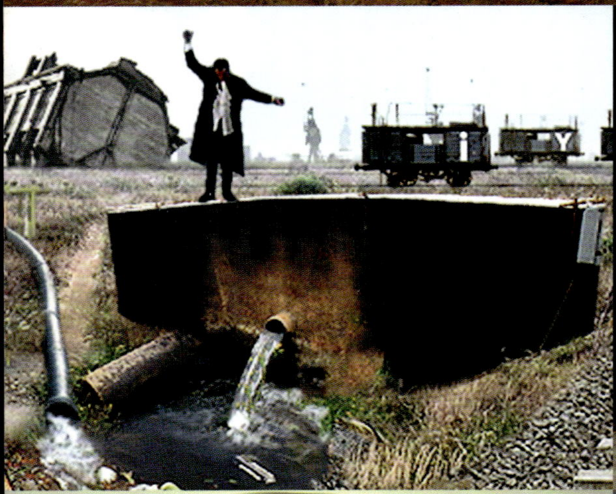

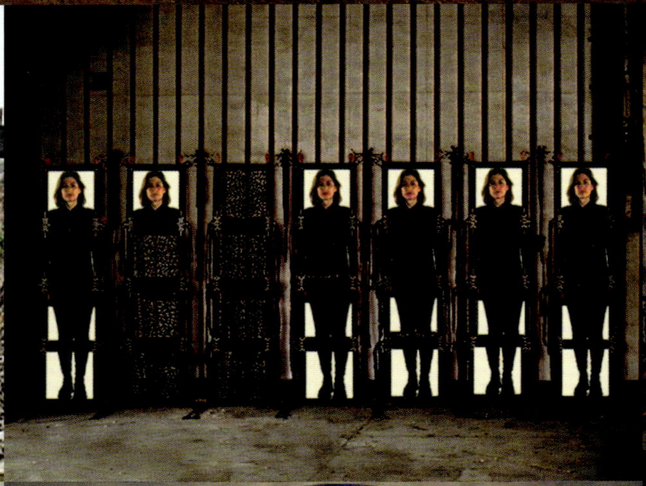

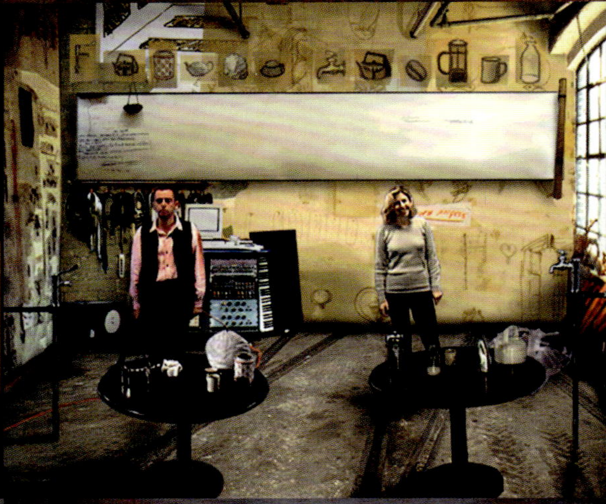

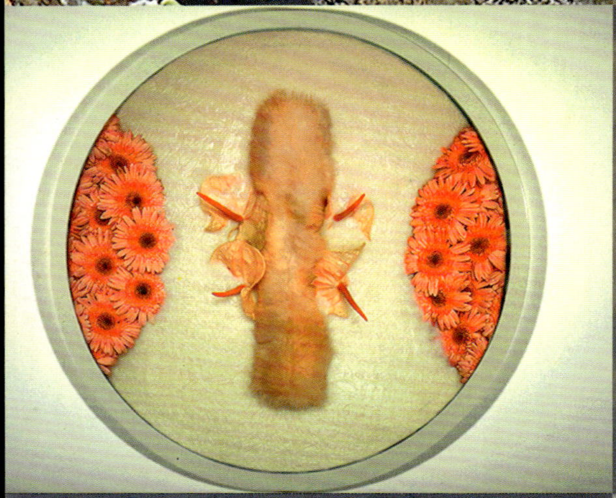

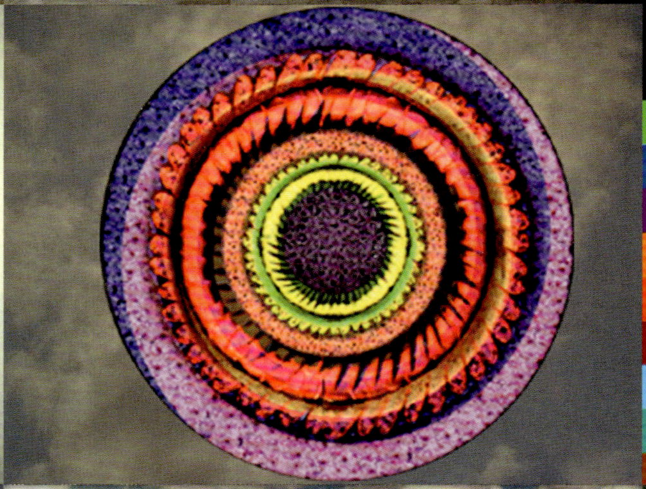

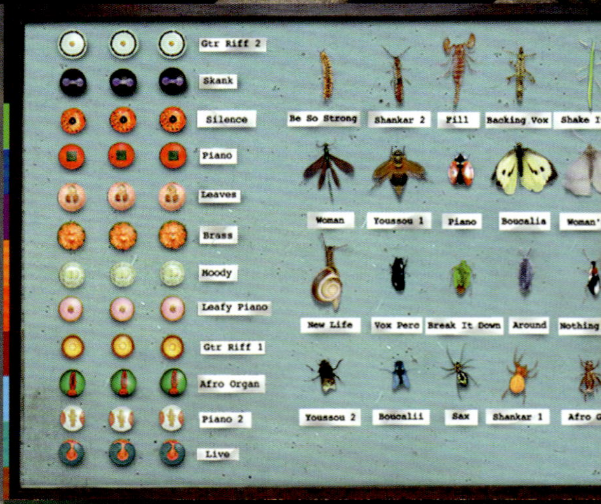
Gtr Riff 2
Skank
Silence
Piano
Leaves
Brass
Moody
Leafy Piano
Gtr Riff 1
Afro Organ
Piano 2
Live
Be So Strong
Shankar 2
Fill
Backing Vox
Woman
Youssou 1
Piano
Boucalia
New Life
Vox Perc
Break It Down
Around
Nothing
Youssou 2
Boucalii
Sax
Shankar 1

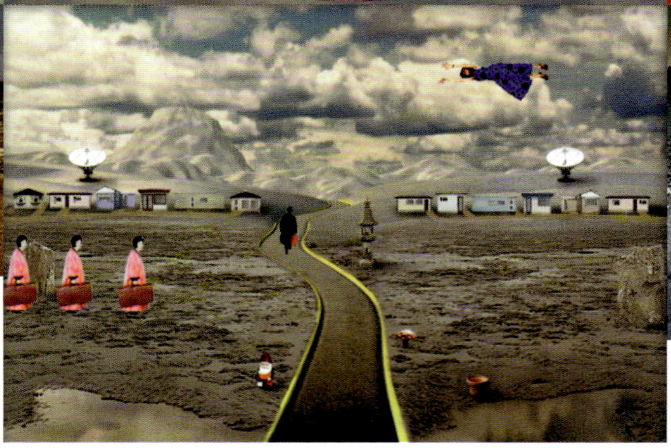

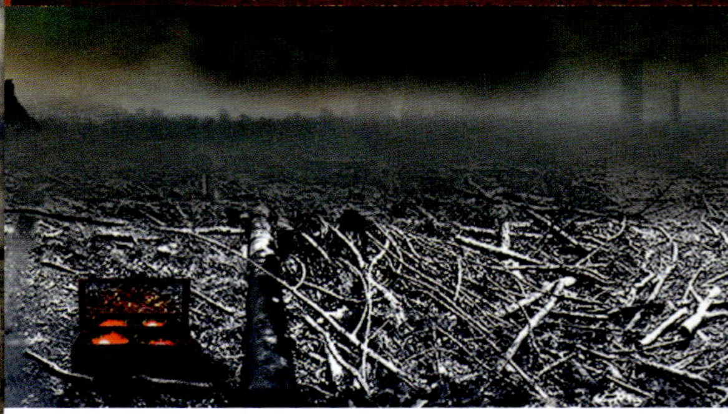

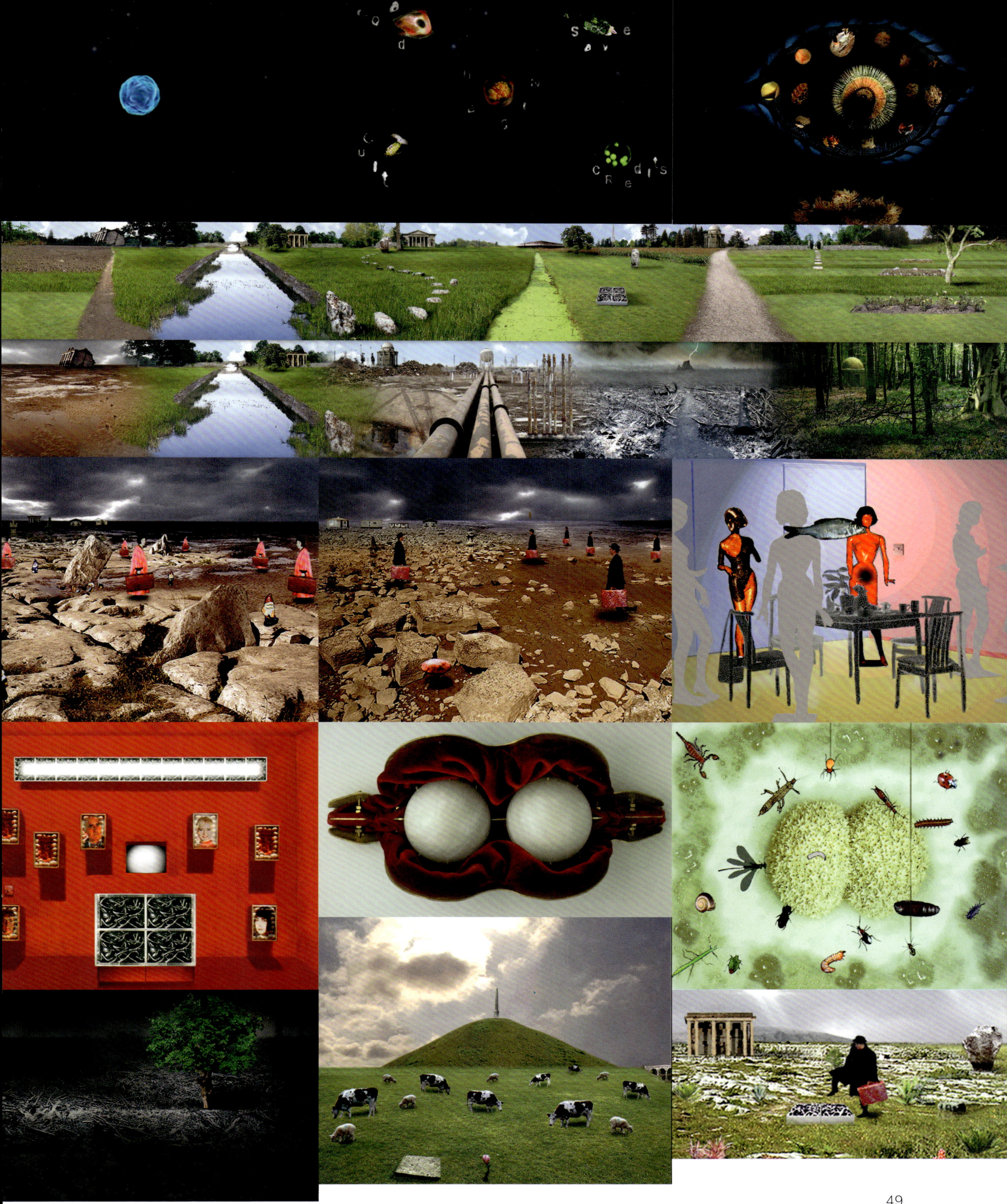

TWELVE STORIES

WERNER HAMMERSTINGL

Story 1

Unexpected discovery.

During a visit of an ageing relative, 'Ethics', who was whiling away his remaining time on this planet, reminiscing about the good old times, our writer accidentally discovered 'Art'. Art, by himself, in a clean white cube-style room, looked old. Appearing fatigued and frail, Art was asleep, alone in the room and hooked up to a full life-support system.

After all the recent claims of Art sightings in various locations around the world, made by critics and academics and the odd citizen (who wouldn't recognise Art if he fell on them), it seems peculiar that Art was discovered in such an inactive state.

Obviously someone is interested in keeping Art alive, despite his obvious frailty and the unlikeliness that he will recover the vigour that we all remember so fondly.

Who?

Story 2

Ingres did the entire painting from sketches.

Data projector lashes into action. The lecturer quietly converses with his computer: yes, the software has everything ready. All the URLs are pre-loaded into the presentation panel. Netscape 24.1 is sooo intuitive . . .

Before the the words 'Jean Auguste Dominique' have completely rolled off his tongue, before the badly pronounced 'Ingres' has a chance to follow, the computer has already anticipated and created the necessary links.

The lecturer makes references to various sketches and demonstrates how they convincingly make up most of the figurative content of the circular canvas.

The students are mostly dozing. If they need any of the information presented here, they will get it from their laptops. Nobody has used the word 'plagiarism' for years.

Some drawings were executed several years before 1852 when the painting itself began.

The Turkish Bath no longer hangs in the Louvre.

Why?

Story 3

Interlace the Futurist manifesto with OS error message listing. Pick colours that also work well in Grayscale.

Will this work?

Story 4

Disembodied intellects proliferate on the interface. They compete for primary positions. This means the retention of the margin.

How?

Story 5

Imagine the memes of the entire aesthetic dimension of our time suffering from a variant of Huntington's disease.

Can you?

Story 6

Can 'Art' take on a new name and identity, like someone in a witness protection programme; start again . . . someplace else . . ?

Can he?

Story 7

Digital Aesthetics.

Yes, I know about that. I use a Holbein Number 5 brush, having of course rounded the pig-bristles a little more away from the flattish middle of the curving tip by lovingly re-scripting that part of the brush. The Shaft of the brush was moulded to my fingers by a Korean custom stylist. My avatar does all my work. I make tiny 'likenesses' in the style of Daguerre. Precious, stilted and restrained likenesses for my publicity-shy clients. They pay me well.

For what?

Story 8

Something about passion.

'After all this time, we still don't know where passion goes when it goes.' This sentence remained after all the other words from the book were forgotten. I think it was Tom Robins that typed them first for 'Still-life with Woodpecker'. I wanted to write something about passion and all I could come up with was a quotation, a drifting fragment. I am learning more from what I cannot do, from what I could never do, or from the unlearned skills that are barely a flicker of information in the now barren expanse of my memories.

Who was I before?

Story 9

Something about truth.

We had just walked through a romantic and awe-inspiring frozen gorge. Wet and clammy granite around us, natural drama unfolding in the form of water rushing past bends, which would slow it enough to cause it to freeze. The Hoelentalklamm near Garmish-Partenkirchen is beyond Wagnerian. I was eight or ten. The whole family was on one of its rare outings. My mother had recently purchased an Agfa camera. It was chrome and warm grey, with a grey vinyl case and strap. She lovingly allowed me to carry this, the most advanced and precious technology in our household, through the gorge. Naturally I was not allowed to view the damp and dangerous beauty of the gorge through the camera or take it off my shoulder for that matter. It dropped. I could feel the sudden limpness of the loose strap sliding off my shoulder. The impact of the camera on granite produced a clatter that terrifyingly pre-empted my mother's hysterical reaction.

Last Christmas my son, aged 12, gave my mother a present. He plays politics better than his father so he produced a card to go with it. I mean he made the card himself. How? He scanned one of Larson's cartoons, one he liked, into Photoshop, positioned the image, coloured some parts and printed it with one side blank to allow for folding and captioning. He wrote some standard Christmas greeting, but personalised for my mother, (it included her name), into the blank half.

She exclaimed with superlatives about the fact that he made this card for her and that 'he is so clever', knowing how to use a computer and 'that she could never do anything like that'.

These two events will always mean more to me than the sum of their parts.

Why?

Story 10

Something about expectations.

We mine the future for elements of truth that we are not yet able to contradict.

Does certainty provide comfort?

Story 11

Forgetting.

Acid on a Rembrandt, a virus in the software, grass finally growing on a mass-grave.

Upset ?

Story 12

Remembering.

The first computer I used was a Tectronics desktop. In 1978 it felt important to sit in front of this machine and make myself familiar with the code named 'basic' (learning quickly that 'basic' and 'simple' actually mean quite different things). That same year I learned much about Dada and Surrealism. Marcel Duchamp became my hero. I worshipped his mind, his clarity and understanding, and insatiably read everything I could find on his life and work. There was a lot. I still have a shelf full of books on Duchamp and I still admire his mind.

The Techtronics machine had cost about three times the price of a new family car or half the value of a house. It had less intelligence than a family pet, but roughly the same capacity of built-in memory.

My analogy with the Techtronics and the family pet does not stop here: the similarities continue when I remember my bloody-minded all-nighters working on a programme for 'hidden line removal'. It is in the same category as trying to make a cat eat something it prefers not to. Even if you succeed, you have had to endure more than the corresponding value of outcome.

Duchamp understood our craving for empiricism, absolutes, truth and value. That's why he manoeuvered these expectations into a check-mated position so frequently and so easily.

Are we having fun yet?

Ending.

COMPUTER POWER

MICHAEL GUIDA

Computer power is real. Digital machines can store and retrieve huge amounts of information, they display lightning-fast reactions and are relied upon to carry out important operations throughout the modern world. But what are the sources of computer power? Do they stem from the machine itself, or from elsewhere?

The Universal Machine

The computer was built to solve mathematical problems, but it was soon realised that its power went beyond numerical calculation. It could manipulate arbitrary symbols as easily as it could add numbers. Lady Lovelace, the disciple of Charles Babbage, understood the power of machine computation in the last century when she remarked that the Analytical Engine, 'can arrange and combine its numerical quantities exactly as if they were letters or other general symbols; and in fact it might bring out its results in algebraic notation, were provisions made accordingly'.[1] From this grew the notion of a computational machine with potentially universal applications.

It is this universality that can make it hard to locate computer power. Joseph Weizenbaum, in his book *Computer Power and Human Reason* (1976), claimed that it is difficult for two reasons: first, that the uses of computers are almost infinitely various, and second that computers are commonly integrated into the structure of other machines.

But, as Alan Turing demonstrated in 1936, the computer can be called a universal machine because it can follow any rule-based procedure that is expressed in an unambiguous and precise language. All processes that can be broken down into a set of procedures can be operated by the regulatory engine of the computer. Stemming from this concept is the claim that a computer can do almost anything. Since all machines have inbuilt limitations, this seems to be an exaggeration, but with time (and often very little time) those limitations are overcome and progress marches on again.

Computers place an inordinate emphasis on the technical process of communication, but somehow offer little in the way of substance themselves. With the exception of the electric light, there has never been a technology that better demonstrates Marshall McLuhan's slogan 'The medium is the message'. The computer is almost all process, directing rather than performing work of its own. The technology of 'command and control', as Norbert Wiener has called it, is of little value without something to control – generally other machines. The space shuttle, for example, is almost entirely under the control of hugely powerful computers, but unless the rocket engines produce enough thrust to launch it there is no mission for the computers to control. Jay David Bolter believes that the subtlety of the computer as a machine dissipates some of the prestige of the older power machines like steam and rocket engines: no longer agents in their own right, they must now submit to the controlling influence of the computer.

Bolter has described the computer as the defining technology of our age. It is the contemporary analogue of the clocks and steam engines of the previous centuries; as important to us as the potter's wheel was to the ancient world. It is not that we are unable live without computers, but that we are different people because we live with them. Bolter suggests that all devices have the potential to become defining technologies because all to some degree redefine our relationship to nature.

Until recently, our technical skills were so feeble compared with the natural forces of the world that we could not seriously affect our environment, except perhaps over the course of millennia. Modern technology, however, promises to give us a new power over nature – our own nature and that of our planet. Of all the elements in our technological future, it is the computer that seems likely to have the greatest impact.

But for a technology to be truly redefining, it must progress from its birthplace to a higher state within the consciousness. In other words, there must be something in the nature of the technology, in its feel, that appeals to the heart as well as the head – for example, the sounds and smooth motion of a piston engine or the autonomy and intricacy of a pendulum clock.

> Such qualities combine with the social and economic importance of the device to make people think. Very often a device will take on a metaphoric significance and be compared in art and philosophy to some part of the animate or inanimate world. Plato compared the created universe to a spindle, Descartes thought of animals as clockwork mechanisms . . . [2]

Today, the computer is a familiar metaphor for the brain.

As a machine, the computer shares qualities with many of its technological predecessors and yet the combination of these qualities is unique. The old idea of a technology was that which could produce a 'controlled application of power to manipulate the environment'.[3] But the computer does not do this. What it can do is serve as a regulator for other machines and it is the information the computer manipulates that makes it useful. In that respect the computer is more like the clock than the steam engine.

The digital computer as an information processor goes far beyond the clock, however. The processor has no cogs or wheels, it achieves its calculations by manipulating electric currents. It is a smooth-running, silent, motionless machine whose calculations are invisible. These features lend it a sense of mystery and magic. Interestingly, part of the job of human-computer interface design seems to be to hide the inner workings of the machine. The user is asked to skim around on the surface while the machine gets on with its unfathomable work somewhere below. Speed takes on new meaning in the domain of the computer, for calculations are performed at unimaginable rates by whizzing electrons.

Unlike the clock or the steam engine, the computer is not a fixed mechanism. This is a machine that is infinitely flexible – by loading a new programme we create a new machine. The computer can even be programmed to write its own programmes, to 'learn', to change itself.

Humanising The Machine

> The instruments man uses become . . . extensions of his body. Most importantly, man must, in order to operate his instruments skilfully, internalise aspects of them in the form of kinesthetic and perceptual habits. In that sense at least, his instruments become literally part of him and modify him, and thus alter the basis of his affective relationship to himself.[4]

It is often said that the computer is merely a tool. However, many machines can be thought of more accurately as functional additions to the human body – 'prosthetic' devices, as Weizenbaum puts it. Some, like the lever and the pneumatic drill extend the raw muscular power of the operator; others, like the microscope and stethoscope are extensions of our sensory facilities. The spear and the radio permit us to cast our influence over a range greater than that of our reach and voice. Our vehicles allow us to travel faster and further than our legs alone can manage.

It is easy to see how such prosthetic machines have enhanced man's sense of power over the material world: 'they tell man that he can remake himself'.[5] In this way, the machine becomes a special kind of tool – it becomes human. With the computer, the effect is similar but greater still. This machine allows man to communicate over huge distances, to produce music, pictures or text, but most of all it allows us new ways of thinking. It becomes a kind of prosthetic mind where we can play with ideas, rearrange them and look at ourselves in a different way.

Consequently, as N Postman has interestingly observed, machines can eventually inherit the name of the humans who once performed their task. For example, the term 'computer' used to refer to a person who performed some kind of mechanical calculation. (The word 'typewriter' has undergone the same mutation.)

When 'it' Becomes us

The idea of a computer that can be intelligent and powerful enough to serve as judge, doctor or psychotherapist is usually dismissed as unachievable. Some feel threatened and insulted by the suggestion that the subtle complexities of the human mind can somehow be reproduced by a computer. There is an urge to draw a line between the natural and the artificial: computers do not have feelings, do not experience pleasure or pain. But when faced with a machine that displays any type of 'intelligence' we find ourselves strangely compelled to treat it as if it were human. In doing so we perpetuate the myth of the knowing, powerful machine.

When computer systems go wrong we are happy to blame them. Even those who know better can experience the computer as a culpable autonomous entity. On the morning of completing this text, two news stories featured computers as guilty of failure: the Ariane space rocket launch apparently failed through a 'computer programme' error (not a programming or programmer error, of course); and British Gas' billing system had gone haywire because of a malicious computer virus.

According to Sherry Turkle, computers tend to be anthropomorphised because of their irreducibility. Describing what a computer actually does is very difficult, but if we don't say what it really does then we talk of it 'thinking'. The problem is that the computer is not like anything else in any simple sense. Finding analogies for what planes do is easy enough – they fly like birds. In the case of the computer, it seems there are no easy analogies, except, of course, with humans.

People usually distance themselves from computers by talking of the distinguishing human qualities that computers can never have – emotion, consciousness, free will and common-sense. But the very lack of these qualities makes the computer all the more powerful. Computers, in their unrelenting and logical way, only do what they are told. They are not susceptible to the stresses or distractions of life. They don't have bad days, except for the rare occasions when they break down.

The metaphor of the machine as human is sufficiently powerful to have made a real impact on everyday language. After the first computer 'virus' in 1988, newspapers noted that computers were 'infected' and that the virus was 'virulent' and 'contagious', that attempts were being made to 'sterilise' the network where the virus had appeared and that programmers were hoping to develop a 'vaccine' against new attacks.[6]

This picturesque anthropomorphism reflects a new relationship between humans and computers. With just one further leap of faith, it can be assumed that the computer has reasons of its own for behaving in a certain way. So when the bank is unable to supply an account balance because the 'computer is down', the implication seems to be that nobody at the bank is responsible: the computer has had a lapse of some sort. Stanley Milgram, in his research on obedience to authority, calls this process 'agentic shift'.[7]

> Because of its seeming intelligence and impartiality, a computer has an almost magical tendency to direct attention away from the people in charge of bureaucratic functions and toward itself, as if the computer were the true source of authority.[8]

Human obedience to the authority of the computer is often startling. Statements like 'the computer shows . . .' or, 'our program has calculated that . . .' are almost always accepted as a preface to a statement of truth. Such statements are akin to saying, 'it is God's will'. Questions about who set the specific parameters for processing the data or who put the data into the computer are rarely asked. This myth that machines, particularly computers, are infallible has a history.

In the early 19th century, the exhibition of what was claimed to be a chess-playing automaton created a sensation in the USA. The machine was a fake. Inside the chess-playing machine was a chess-playing man. Edgar Allan Poe saw the machine in action and wrote about the experience. His explanation of why the machine was a sham reflects the popular image of the machine at that time, which still has currency today. For Poe, the machine was proved to be bogus because it lost a game. 'The Automaton does not invariably win the game', Poe explains. 'Were the machine a pure machine, this would not be the case – it would always win.'[9]

'Machines, when they operate properly, are not merely law abiding; they are embodiments of the law', claimed Weizenbaum. We expect an ordinary desk calculator to be an embodiment of the laws of arithmetic. Should it deliver an unexpected answer, we do not question the laws of arithmetic nor the laws of the machine. Rather, we assume that we have made an error when punching in the data.

If computers work simply by reading rule-based instructions then in that limited sense they are infallible. They do exactly what they are made to do. And perhaps, following Weizenbaum, this is why the public so happily embraces the notion of computer power despite only the slightest understanding of what might be happening inside the machine. The public vaguely understands – but is nonetheless firmly convinced – that any rule-based procedure can theoretically be carried out by a computer. Since man, nature and society carry out procedures that are surely rule-based in one way or another, it follows that a computer can at least imitate them in all their procedural aspects. 'Hence, everything . . . is at least potentially understandable in terms of computer models and metaphors.'[10] That is, computers can do anything.

Programs for Artificial Intelligence are not scientific constructions, but actually wonderfully clever ornaments. But we should not underestimate the value of such decorations. For example, as Bolter has pointed out, the crowing cock and the animated figures on the Strasbourg clock of the 14th century captured the contemporary imagination more readily than the concealed mechanism that actually told the time. The Artificial Intelligence movement is important because it produces programs that illustrate as much as perform. The power of these applications is not that they may have practical uses but something altogether more marvellous – they demonstrate what it means to live in the computer age. Achievements such as a chess program that can take on the world's best minds create a sense of awe in the public.

Computer memory is like human memory: it is multidimensional and associative, can possess a networked structure, and has a capacity for change: computers are always ready to replace old memory with new, modify or erasing existing data.

Part of the power of the computer comes from its ability to store and retrieve data reliably, accurately and at high speed. It is probably true that all of the information being generated by today's society cannot be mastered except by electronic means. The computer allows information to be corralled and structured, placed in databases and preserved so that it can be found again easily.

Such mastery is probably as important to us today as it was in the Renaissance: memory being the key to human knowledge and therefore to human control of the world.[11] The scholars, Bruno and Camillo constructed memory devices (a memory wheel and a memory theatre, respectively) to reflect the true structure of the world and all humanly accessible knowledge. To be able to manipulate these devices was to be able to win that knowledge and through magic establish a rhythm with the world. Although the computer programmer does not think of his programs as supernatural, he does think of them as accurately reflecting the world outside his computer.[12]

A computer database is the modern equivalent of Bruno's memory wheel: its purpose is to allow the user control over far more information than could ever be kept in the head. Within databases lies power. Actually, what happens in the production of large databases is that the selection and 'memorisation' process is carried out by ranks of cataloguers or typists who input the information. Then the machine can work its magic, analysing, re-presenting and revealing structural relationships within the data that the human eye could not easily perceive.

The World Wide Web is now a vast database of information from around the globe. A range of search engines allow the user to select from the overwhelming array of data that lives on computer servers. Currently, the search engines are somewhat unreliable and information retrieval is slow. But a huge interconnected web of data, cross-referenced and hypertextually linked is a truly powerful ideal.

Since the 1960s, specialists have been talking about the possibility of storing all recorded human knowledge in one gigantic database. The CYC project in Texas seeks to produce a machine intelligence with common sense.[13] For ten years, millions of bits of information have been tapped into the program along with tens of thousands of rules. CYC will be more than a database of facts but at its core is a huge electronic knowledge base. All governments and corporations rely on knowlege bases of market trends, client preferences etc. It would seem that these electronic systems are heirs to French encyclopaedias of the 18th century and to Camillo's memory theatre of the 16th century. They are the electronic embodiment of the conviction that memory is knowledge and knowledge is power.

Identification with the Machine

Referring to ourselves as mechanical devices or computers is second nature. In a recent television programme, a woman said she felt her brain had become 'unplugged' and 'short circuited'. People now commonly speak of 'programming' and 'reprogramming' themselves.

One can detect the same metaphor in medicine. Doctors and patients have become accustomed to the idea that the body is made up of parts that, when defective, can be replaced with bits from other humans or synthetic substitutes without affecting the rest of the body. To date, the brain has been spared this mechanistic approach.

However, the comparison of the mind and the computer is almost irresistible. The analogy is all the more compelling because we know that our brains are made up of neurones that transmit information by tiny electrical signals. In a simple sense this is exactly what the computer processor does. The AI community has spent a great deal of energy trying to map the mind onto software and create a thinking machine.

But this is not a discussion of whether computers can really think or not. It is about whether people believe that there are thought processes within the computer's box, and whether humans are able and willing to take on the qualities of digital computers. Turkle believes that the real unstated question lying behind our preoccupation with the capability of computers is not what will they be able to do, but what will they do to us? What kind of people are we becoming? Computers challenge our ideas about what it is to be human, to think, to feel. Computers 'present us with more than a challenge', she claims, 'They present us with an affront, because they hold up a new mirror in which mind is reflected as machine'. To her the computer is the 'first psychological machine'.

> Some objects, and in our time the computer is preeminent among them, provoke reflection on fundamentals. Children playing with toys that they imagine to be alive and adults playing with the idea of the mind as a program, are both drawn by the computer's ability to provoke and to color self-reflection. The computer is a 'metaphysical machine', a 'psychological machine', not just because it might be said to have a psychology, but because it influences how we think about our own.[14]

As computers begin to display human characteristics, some people respond by underlining the differences between the human mind and the machine. Others relish the similarities. Weizenbaum in *Computer Power and Human Reason* argues that the computer – logical and rule-governed – encourages us to differentiate ourselves from the machine. He fears that the psychological theories of Artificial Intelligence would lead to a flattened, mechanical view of human nature. His response is to value as most human that which is most different from the computer. As emblematic of the human essence, he chooses what the computer cannot do: 'the things we know but cannot tell'; the glance worldlessly shared by a mother and father as they stand over the bed of their sleeping child.

Resistance to a computational model of people as programmed information systems leads to Weizenbaum's view that what is essential in the human is uncodable. When we use information-processing models to explain larger and larger slices of our behaviour, we seem impelled to isolate as our core something we can think of as beyond information. This idea is echoed by the philosopher John Searle who argues that no matter how perfect a computer, simulation of thought is not thought, because the computer will simply be following rules it does not understand.[15]

Searle provides an antidote and a challenge to the florid claims of the AI community. When people subscribe to his argument they are doing much more than merely agreeing with his view. They are using Searle to help themselves reassert a set of deeply held beliefs: a commitment to the self, to the 'real', to the human as unique.[16]

For Searle, thought is the product of our specific biology, of a human brain. And biology brings with it vulnerability, sexuality and mortality. This expression of what is human is not mere romanticism. To make such distinctions is not to reject the computer. It demonstrates what the technology of the computer has done to the way we think about ourselves.

> The computer gives support to those who see human psychology in mechanistic terms, and also, in a paradox that is increasingly important for our culture, it is a point of reference for those who place greatest value not on rationality but on affect. We cede to the computer the power of reason, but at the same time, in defence, our sense of identity becomes increasingly focused on the soul and the spirit of the machine as human.[17]

Man as Information Processor

The title of Bolter's book, *Turing's Man: Western Culture in the Computer Age,* conveys an important idea, similar to what would be suggested by saying that from the 16th century until recently, we were 'Gutenberg's Men'. Bolter argues that the computer is the dominant metaphor of our age – it defines our era by proposing a new relationship to information, work, power, and to nature itself. The computer redefines humans as 'information processors' and nature itself as information to be processed. The fundamental metaphorical message of the computer in short is that we are machines, thinking machines, but machines nevertheless.

Postman believes that the computer is so powerful it subordinates our nature, our biology, our emotions, our spirituality. It claims sovereignty over the whole range of human experience and supports this claim by showing us it can 'think' better than we can. In his enthusiasm for Artificial Intelligence, Marvin Minsky has stated that the thinking power of silicon brains will be so formidable that, 'If we are lucky, they will keep us as pets'.[18]

Faking it – Artificial Intelligence

The computer as a thinking machine that rivals or surpasses human intelligence, is a familiar theme in computer science and in science fiction. Ever since Alan Turing published his polemic *Computing Machinery and Intelligence*, in 1950, stating that computers were capable of imitating human intelligence perfectly and would do so by the year 2,000, the idea of the thinking machine has captured the imagination of scientists, philosophers and the public. What could be more powerful than a thinking machine?

In fact, the success of computers in so many domains has led to the widespread assumption that they already possess human intelligence. We could blame HAL in Stanley Kubrick's film *2001: A Space Odyssey* and other super-intelligent computers depicted in cinema, but these are of course symptoms rather than causes.

Our culture seems eager for, or resigned to accept, computers that can talk, read the newspaper, control traffic and make economic and political decisions for us. In fact, we assume computers are already busy doing these things. But the reality is that they cannot. That is, the computer can perform none of these tasks autonomously although it can play a role in all of them through interaction with a human user. Artificial Intelligence, however, insists on the autonomy of the machine: the goal is to embody in the computer an intelligence that can operate in isolation of human users.

Since all the clamour about Artificial Intelligence began in the 1960s very little has actually been achieved. But strangely, this does not seem to have diminished our interest in the potential of the intelligent machine. The idea of a computer carrying out seemingly clever processes is enough to sustain us. We are compelled by the notion of duplicating the human mind or at least aspects of it. Whether computers can truly be said to think, and whether the brain is just a digital computer and the mind simply a computer program are perhaps best left to the philosophers and Artificial Intelligentsia to discuss. With the conversational program, 'Eliza' (1976), Weizenbaum has demonstrated that a computer does not actually have to think, only to give the illusion of thought and intelligence. A computer program that simulates rather than duplicates intelligence may not be enough to satisfy the AI specialist, but it is often enough to convince others. We are quite happy to project intelligence on a system, seemingly for our own amusement.

Inverting the original Turing test (the 'Imitation Game'), 'Eliza' asks the questions and the human reader answers. If anything, this makes the task of impersonating a human easier, because the subject has already ceded control to the program, which takes on the persona of a psychotherapist whose questions are all rhetorical. Weizenbaum was startled by how quickly and deeply people conversing with 'Eliza' became emotionally involved and how unequivocally they anthropomorphised it:

> People who knew very well that they were conversing with a machine soon forgot that fact, just as theatre goers, in the grip of suspended disbelief, soon forget that the action they are witnessing is not 'real'. This illusion was especially strong and most tenaciously clung to among people who knew little or nothing about computers. They would often demand to be permitted to converse with the system in private, and would, after conversing with it for some time, insist, in spite of my explanations, that the machine really understood them.[19]

In fact, 'Eliza' is constructed from a series of tricks that constantly turn the users' words around. The program focuses on key words and repeats fragments of recent responses. Ultimately, what happens is a monologue, in which the human being is unwittingly responsible for both questions and responses. The fact that the experience is based on a trick is acceptable as long as the user is seduced by it.

Playing with Power

The enthusiasm for Artificial Intelligence stems in part from the age-old desire to recreate life through technology, the desire expressed in the myth of Pygmalion, in the legend of the Golem and in the alchemists' formula. 'Artificial Intelligence is the newest recipe for the homunculus', as Bolter puts it.

Seeking to create a sense of life within the realm of the computer, I produced a lie-detector machine with a colleague, Mark Winstanley. It comprised simple 'Yes' and 'No' input buttons and a 'galvanometer' device in which the right hand was entrapped. The subject was asked a variable series of about 45 questions by the computer. The work is known as 'Sex, Lies and Binary Logic', and was shown in the UK and France.

We drew heavily on the essence of the computer, not so much as an 'analytical engine', but as an evocative object, with its attendant mystery and invisible complexity. Its rational behaviour was its personality. The computer was given a sinister and authoritative persona and spoke with a synthetic voice, actually a hoarse whisper.

In the same way that we assume a desk calculator follows the laws of arithmetic, can we assume that a lie-detector machine embodies the laws of truth? And does the computer embody the laws of reason? These questions underpinned our thinking and we played on them as we created the personality of the machine (its speech characteristics, the visualisation of its thinking on screen) and built up the mythology of what a lie-detecting device might be capable of.

Our intention was to contrive a reversal of the roles of the subject and the computer. In other words, the subject was forced to submit to the binary world of yes/no, true/false, while the computer was empowered through speech, thought, decision-making and detection. The Turing test, 'Eliza', 'Julia' (a program that masquerades as a person on the Internet) and other conversational programs rely on reading and writing to communicate with the user. We found that the subjects felt speech to be engaging, especially as the voice, whilst synthesised, sounded human to many. Like 'Eliza', our program used the computer's responses (in this case, statements and further questions about the subject) to direct the subject/listener back to his or her own resources in a quest for meaning.

During the interrogation the subject, it seemed, brought to the exchange an image of the questioner. All sorts of connections and assumptions were constructed to enable them to turn the experience into a dialogue with a real person. Many subjects earnestly reported intelligent behaviour of the system that had not been built in by us. They imputed to the system their own ideas of what a lie-detecting machine might do.

Like Turing's test, and many so-called 'intelligent' systems, 'Sex, Lies and Binary Logic' was partly a trick; or rather it was designed to be seen as a small piece of theatre. Only the mythology of the lie-detector was in operation. A few small hints of knowledge by the system (knowing the day and the time) plus a couple of occasions of 'smart' feedback to the subject (revisiting a specific answer, for instance) went a very long way indeed. People who took it as a performance seemed to enjoy being tricked as someone might enjoy a magic show. They allowed themselves to be seduced by the power-play that ensued, and submitted to the authority of the computer.

Those who could not or would not suspend their disbelief felt conned or cheated. The lie-detector didn't work, the computer wasn't doing anything clever, they said. Which is partly true – as soon as we stop our worshipful reverence of the machine it becomes the dumb servant it always was.

Computer power comes from us.

Notes

1 Morrison and Morrison in JD Bolter *Turing's Man – Western Culture in the Computer Age*, Duckworth, London, 1984, p66.

2 *Ibid* p11.

3 *Ibid* p45.

4 J Weizenbaum, *Computer Power and Human Reason*, Penguin Books, London, 1976, p9.

5 *Ibid* p28.

6 Gozzi in N Postman, *Technopoly: The Surrender of Culture to Technology*, Vintage Books, New York, 1992, p113.

7 Milgram in RL Atkinson, RC Atkinson, ER Hilgard, *Introduction to Psychology,* (eighth ed) Harcourt Brace Jovanovich, New York, 1981, p205.

8 Postman, *Technopoly*, p115.

9 Poe in S Turkle *The Second Self: Computers and the Human Spirit,* Simon & Schuster, New York, 1984, p371.

10 Weizenbaum, *Computer Power,* p157.

11 JD Bolter *Writing Space: The Computer, Hypertext and the History of Writing,* Lawrence Erlbaum Associates, Hillsdale, New Jersey and Hove & London, 1991.

12 *Ibid.*

13 D Lenat, 'Artificial Intelligence' in *Scientific American*, September 1995.

14 Turkle, *The Second Self,* p197.

15 JR Searle, *The Rediscovery of the Mind,* The MIT Press, Cambridge, Massachusetts, 1994.

16 Turkle, *The Second Self.*

17 Turkle, *The Second Self,* p326.

18 Minsky in Postman, *Technopoly*, p111.

19 Weizenbaum, *Computer Power,* p189.

SURVIVAL RESEARCH LABORATORIES

The Search for Sentient Machines

JULIAN BLOM

SRL's shows are loud, violent, physiologically unsettling. There's nothing quite like feeling your eyeballs and kidneys vibrating to the siren call of subsonic frequencies, or your ears screaming at the outer limits of their ability to handle decibels. These performances make you aware of your body in ways that are usually only experienced in war. Just as the battleground is the site for the symbiosis of man and machine at its most extreme, SRL events throw into sharp relief our relationship with machine technology.

We like to look upon machines as benign children whom we have created to serve us. We perceive that we have control over our technological progeny whom we can safely manage and maintain. Only when we are faced with something like a car crash do we suddenly become aware of the other face of machines; the softness of our bodies is suddenly brought to light by the hard carapace of the car being crushed around us.

In horror films our fear of machines comes to the surface. Here they develop personalities and behaviours of their own. In Fritz Lang's *Metropolis* (1926), for example, the heroine is replaced by a malevolent robot. In Donald Cammell's *Demon Seed*, (1978) a woman is raped and impregnated by a computer. The 'Replicants' in Ridley Scott's *Bladerunner* (1982), on the other hand, are deemed evil when they develop and manifest their own human-like desires – when they become machines in control of themselves. And in *Star Trek: The Next Generation,* the android Commander Data is somehow deficient, always at a disadvantage in his inability to be human.

Does our relationship with our machine progeny really have to be this way? Aren't machines in fact a pure distillation of what we are as a species: our intellectual children? Are we not essentially defined by our machinery? These are the questions that Mark Pauline and SRL answer by making the machines that populate their performances.

A few years ago Pauline was asked by a journalist from Mondo 2000 to explain his motivation for building them. His answer was something like: 'I just look at a piece of machinery or a device, and ask it what it would really like to be doing'.

The result of this process is that a wide range of machines are encouraged to 'become', to perform their preferred function rather than their designed one. For example, over the last few years SRL's shows have featured a Ramjet originally designed to power Hitler's V1s during World War II. Its new role is as a sound generator capable of producing a range of subsonic and audible frequencies that can vibrate every organ in your body.

In 1991 I was in Barcelona at the Art Futura 91 exhibition where SRL were to provide the finale to the week-long Cybermedia event. The first time they fired up their Ramjet, people within a 1km radius of the test ground were convinced that there was an earthquake in progress . . . the fire brigade and police were called out . . . the ground vibrated beneath one's feet.

SRL's huge metallic monsters are the dinosaurs of our time. Creations of our machine technology, they have the same ability as a Tyranosaurus Rex to make the earth tremble beneath our feet, a sensation not heard but felt in our viscera. They come together like beasts in combat, flailing at each other with giant maces or articulated arms and 12-inch, knife-like fingers. Belching flames or condensing the air into concussion waves that burst over your body like hazy smoke rings they gleefully destroy the set.

As the show strips away the thin veneer of civilisation – our controlled and habitual response to our manufactured environment – the machines begin to show signs of inherent behavioural patterns, suggesting to the audience that a new phylum of creatures or alien species is being born in its midst with SRL as midwives. An alien species born of ourselves.

Bearing in mind that organisations such as NASA and MIT cite SRL as the inspiration for some of their functional, autonomous machines, this is not so far-fetched. Rod Brooks at MIT points to SRL's work as one of the sources for his MOBOTs experiments. Small, cheap robots, about the size of a child's radio control car, they possess simple visual abilities and basic mobility, and about the same intelligence as insects. Cheap to produce in quantity, they are designed to be sent to Mars as a swarm of sampling robots, sending back data to earth until they break down.

These days, SRL's creations are becoming increasingly hi-tech. With the end of the Cold War, the US government is selling off a good deal of declassified military equipment for scrap prices – missiles with their brains taken out, jet engines, telemetry systems – and SRL are taking advantage of the glut of goodies. It is Pauline's wish to bring a new intelligent machine species into the world. In 1995, in BBC2's film *Pandaemonium*, he commented: 'I hope that when [intelligent machines] come along, they'll take kindly to the fact that we helped them along.'

Let's hope so!

KENJIE YANOBE

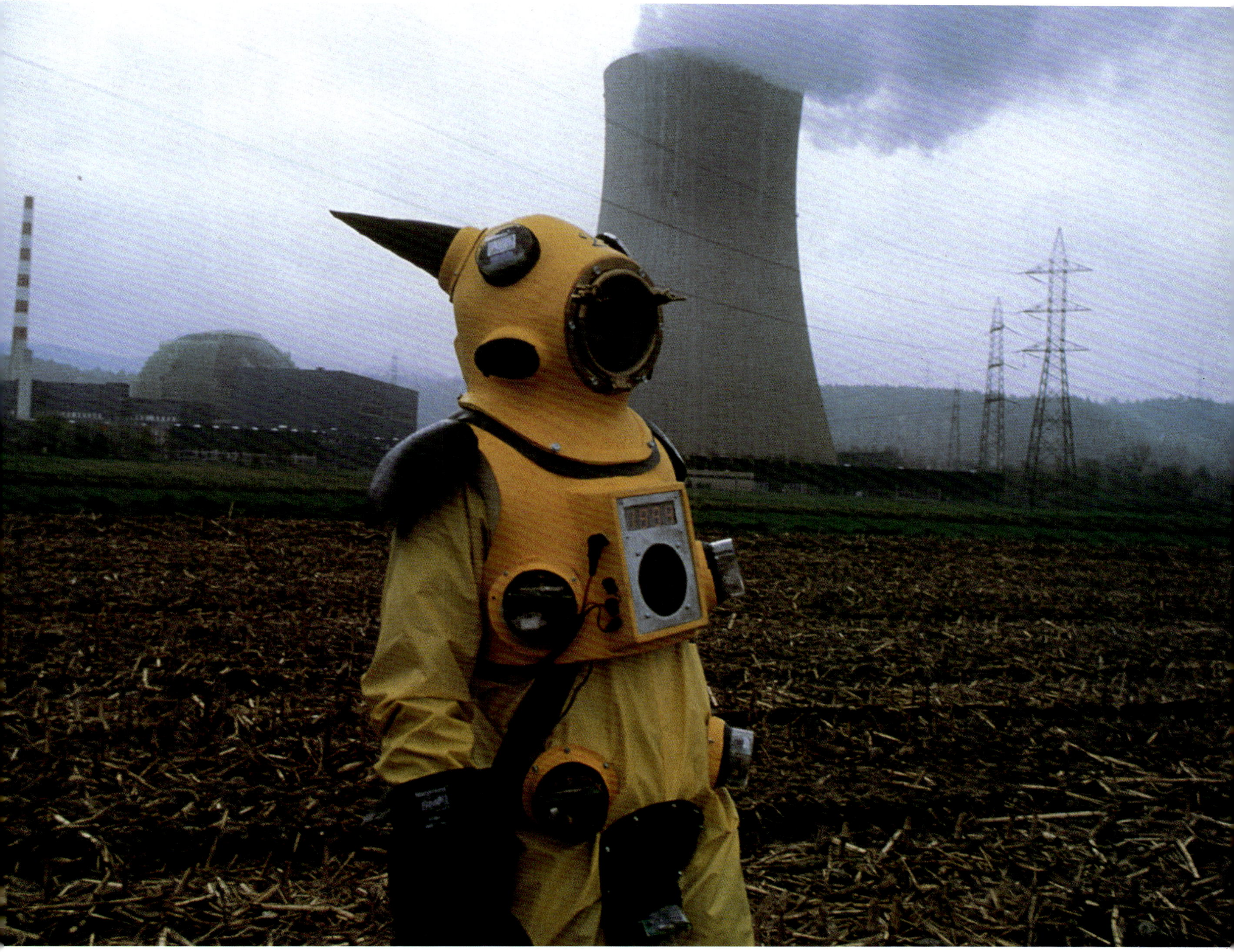

Like many Japanese men of his generation, growing up in a society obsessed with technology and Manga comics, Kenji Yanobe, has been described as otaku, *or 'computer nerd'. His bio-mechanical creations embody a paranoid but humorous view of the future of humankind. Works such as* Atom Suit No 2, *(1996, geiger counter, stroboscop, 180 x 80 x 40cm) are a combination of survival technology and robotics, designed to protect against a future hostile environment. Totally encasing the body, these living sculptures often include plants or aquatic interiors to nourish the intelligent 'wetware' (the redundant body) inside.*

NANOTECHNOLOGY

The Eve of the Diamond Age

JULIAN BLOM

Masahiko reappeared with the thing in his hand minus the duty-free bag. 'I scanned it,' he said. 'Immediate identification as Rodel-van-Erp primary biomolecular programming module C-slash-7A. This is a lab prototype. We are unable to determine its exact legal status, but the production model, C-slash-9E, is Class 1 nanotechnology, proscribed under international law. Japanese law, conviction of illegal possession of Class 1 device carries automatic life sentence.' 'Life?' Chia said. 'Same for thermonuclear device,' he said apologetically, 'poison gas, biological weapon.' He held up the scanned object for Zona's inspection. Zona looked at it. 'Fuck your mother,' she said, her tone of sombre respect. William Gibson *Idoru*

Gibson has it right. The successful development of nanotechnology, or 'molecular' technology, as it is also known, will change the world in ways we cannot yet begin to grasp. It will be as potent a technological breakthrough as the splitting of the atom or the development of bio-weapons, but with the power to change our world and our physical bodies in fundamentally more influential ways, whether benign or pernicious.

Its effect will be greater than that of the Industrial Revolution, biotechnology, genetics and nuclear fission put together. A *tsunami* of change will cascade across the world affecting everything from the water we drink to the cells in our bodies. We are looking at the most pervasive technology yet to be envisaged. It will spread in almost the same way as a virus. And it will be quiet, unnoticeable, until we feel the changes.

In simple terms, nanotechnology is the manipulation of matter at the atomic level. Most manufacturing technologies and scientific disciplines approach the transformation of chemical compounds or solid matter using tools like lathes and milling machines for metals, which shave off micron-thick layers of atoms. Our industrial chemical processes swish vats of chemical components around, allowing nature to bump the molecules together in the hope that a large enough quantity of them will react – a profoundly haphazard and wasteful process that generates volumes of toxic by-products. Though we see ourselves as a technologically advanced species, our production methods are no more than refinements of Stone Age technology. We are still 'forced to handle atoms in unruly herds.' (K Eric Drexler, *Engines of Creation*)

Nanotechnology implies a level of refinement in the treatment of materials that currently defies the imagination. It is the technology of taking individual atoms and using them like Lego blocks to construct objects atom by atom. For example, by applying 'assemblers' – the tools of nanotechnology – to a pile of carbon dust (flexible, durable, cheap and readily available), we could watch it form itself into a car made of diamond.

> Natural diamond is expensive, we can't make it in the shapes we want, and it shatters. Nanotechnology will let us inexpensively make shatterproof diamond (with a structure that might resemble diamond fibers) in exactly the shapes we want. This would let us make a Boeing 747 that would weigh one fiftieth of today's versions without any sacrifice in strength. The benefit to space travel would also be dramatic. The strength-to-weight ratio and the cost of components are critical to the performance and economy of space ships: nanotechnology could improve both of these parameters by about two orders of magnitude. *Dr Ralph Merkle, Xerox Palo Alto Research Centre*

In the development of these 'assemblers', K Eric Drexler, author and pioneer of nanotechnology, has mooted the idea that in order to create technology at this scale, it might be feasible to use variants on nature's own nano-scale tools – enzymes – proteins that are told how to function by small ribbons of RNA (ribonucleic acid). RNA works in much the same way as old computer punch cards, which were encoded with on and off sequences via patterns of holes, interpreted by the computer as a series of commands. Assemblers would be the base tool on which the rest of the nanotechnological universe would be built: tools to create tools to perform any function.

Some two to three years ago, using a variety of scanning electron microscope, IBM and Xerox managed to write the letters 'IBM' on a sliver of silicon. At that point, it was envisaged that it would be at least 30 years before we would see the benefits of nanotechnology in common usage. It seems that these estimates were overly conservative. Nano-engines have been found or created and the first nanotransistor has now been produced in laboratory conditions. The engines to drive potential assembler nanobots and the transistors to create their brains are therefore already on the way.

In order to imagine the effects of this technology, think for a moment of all the things you have fantasised about. Remember when, as a child, you used to wonder what it would be like if you could read another person's mind? Remember the times when you used to feel that sense of connectedness to other people? Remember being so immersed in a book or film that it actually took on dimensionality in your imagination? That feeling that you could melt your body into your lover's and become one being? The desire to be able to see in the dark? The ambition to be more intelligent, more articulate? The wish to save the environment and the human race from pollution? Nanotechnology holds the key to making these things a reality.

Recent developments in nanotechnology include the production of a nano-transistor, only nanometres across (billionths of a metre). This passage from Drexler's book *Engines of Creation*: puts the scale of molecular technology into perspective:

> If an atom were the size of a small marble, a fairly complex molecule would be the size of your fist. This makes a useful mental image, but atoms are really about 1/10,000 the size of bacteria, and bacteria are about 1/10,000 the size of mosquitoes. (An atomic nucleus, however is about 1/100,000 the size of the atom itself; the difference between an atom and its nucleus is the difference between a fire and a nuclear reaction.)

The creation of such a transistor opens the door to a number of astounding possibilities. It can work at frequencies in the PetaHertz range, meaning that molecular computers using nano-scale transistors of this type could function at several orders of magnitude faster than current computers while packing in the corner of a fingernail more processing power than today's fastest computers. This means that, with your body's electrical field providing the power supply, you could have an enormously powerful computer system linked biologically to your neural pathway – that is, your brain would have direct access to this computer, which would exist within your body.

You would then have the ability to collate more strands of information than you ever thought possible. Through high frequency links with other people's inbuilt computers you could communicate with them just by thinking. You could have direct contact with everyone else on the planet, constantly. You could feel the currents of the collective unconscious and be a part of that river of feeling and thought. In essence, the human being would have taken the next evolutionary step.

A cultural anchor illustrating our ambivalence to this kind of technological symbiosis can be found in the BORG, the advanced collective culture imagined in *Star Trek: The Next Generation*. Pallid humanoids with chips and prostheses replacing eyes and hands, the BORG create their food by taking in energy from their cubic ship, where they plug into spaces to form part of the computational matrix of a gigantic bio-mechanical computer collective. The BORG have no emotion, no individuality, no ego; only the drive to 'assimilate' new technologies and cultures. They are in effect the ultimate American villains: techno-communists.

It is interesting that the BORG, who work together without problems and are truly social creatures, should be portrayed as un-American. In their decision to form a symbiotic relationship with technology and by changing the bodies that God gave them, they become anathema. Their pale flesh, studded with prosthetics and chips, is ugly; their lack of ego, sinister. They are deemed less than human. But from their own point of view their desire to 'assimilate' other species and technologies, to add 'the distinctiveness of your species to our own' as they put it, is benign. The paradox here is that we rush at developing empowering, 'godlike' technologies only to revile them.

If we do take responsibility for our own evolution through nanotechnology we will be able to use assemblers to repair our cells and prevent ageing and disease. Our arteries will be scrubbed clean of cholesterol; our blood clots will be broken down before the thrombosis hits our lungs. No more brittle bones; no more clogged lungs; no more congenital disease; no more death! And no external signs that we are any different from before. Unless, that is, you want to breathe under water using your lungs. Then, of course, the assemblers in your system would create you some gills. Or let's say you lose a limb in an accident: the assemblers would grow you a new one. Or perhaps you want to see what it is like to be a member of the opposite sex? Assemblers would perhaps change your genetic code, or just grow you the appropriate physical parts.

In this potential future you could chose your skin colour as a matter of whim. Maybe it would be fun to go into a bar with mirrored skin and over the course of the evening change to black and white with Rorschach ink blots all over your face. Perhaps you would want to get really close to your loved one and cause your hands to bond together, joining nerves at the same time so you share the same sensations. We are on the cusp of becoming a new species; a species with a godlike capacity to evolve to suit its changing environment.

Now we've been off into the glorious and scary future of a world with nanotechnology, let's just take a step back. The changes that this potent technology is capable of producing on the human race and world are unquantifiable. We don't know whether introducing computer technology into our bodies will be beneficial or will turn us collectively insane. Nor do we know if a nanotechnologically enhanced human being would think in a way that we could even begin to comprehend; or whether assemblers used for the purpose of solving one problem will not cause another. The same technology that can offer the possibilities outlined above, could be used to clean up the mess we have made of our world, but could also make it worse. We only need to look at previous miracles of science and technology – DDT, Thalidomide, biological weapons, nuclear power – to see how this can happen. We should pause and ask the question: just because we *can* create this technology, does that mean we should?

PATRICE CAIRE

C.A.I.R.E.'s Cyberhead

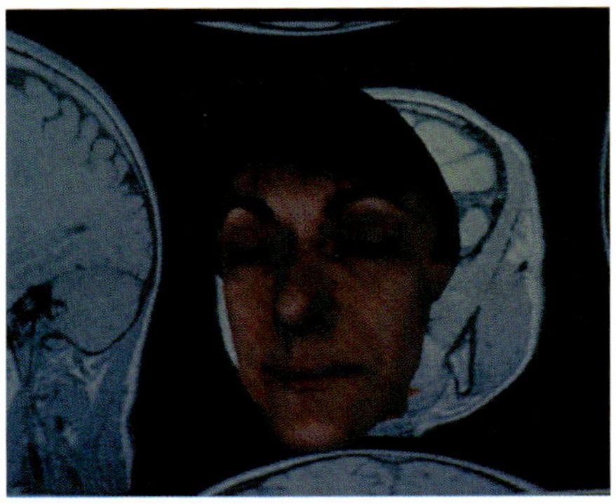

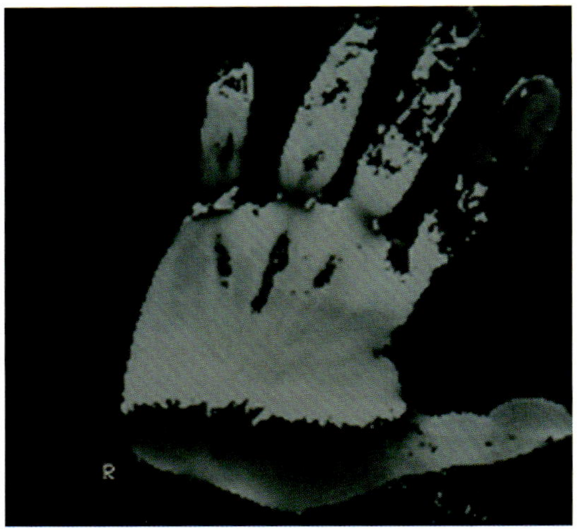

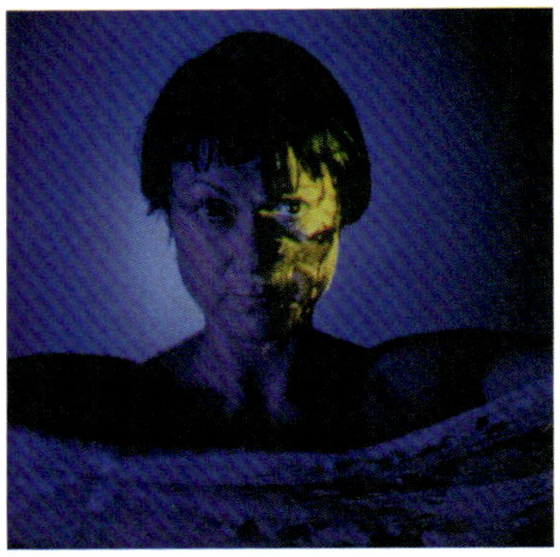

> *A drop of quicksilver spills onto the walls of the darkened gallery. This liquid spreads, blossoming into brains and bone to form a fleeting humanoid profile which instantly melts and disappears.* 'Cyber Art's Inner Reaches Exposed' C.A.I.R.E., by Erin Blackwell, *Bay Area Reporter*, June 1994.

Patrice Caire is an artist who employs innovative technologies to create new links between science, philosophy and art in order to expand cultural boundaries. She utilises artificial realities as a medium of expression and experience as well as a new way for people to interact with one another:

> My installations are micro world learning environments: I create situations to be shared, and my objects are space and data. I use artificial realities as a medium of expression and experience, as well as a new way for people to interact with each other. These artificial realities emerge out of a new way of thinking about computers and their relationship to human experience. In contrast to the dominant model of computer interaction, based on nodes and links expressed as binary realities, my models are created from the notion of connection and weighted links. The data sources are based in reality whilst allowing the viewer to experience a possible reality but not necessarily an actual one.

Her installation *C.A.I.R.E. Cyberhead-Am-I-Really-Existing* (1994) is an absorbing experience in which the audience is invited to navigate through an electromagnetic portrait of the artist's head. Caire believes that our reality is shaped by the patterns of our bodily movements, the contours of our spatial and temporal orientation and the forms of our interaction with objects. She uses the analogy between anatomical structure and subjective experience as the metaphorical and literal basis for creating a highly personal Virtual Reality experience that enables the audience to engage in her thought processes.

During interaction with *Cyberhead*, the user becomes a necessary component of cyberspace – a product of artificial experience. This seamless merging of human imagination, computer hardware and dataspace, has been popularised in the science-fiction mythologies offered by cyberpunk authors.

In William Gibson's cult novel *Neuromancer*, the fluid interfacing between these different realities is explored by the hero, Case. In a future world, dependent on a badly organised computer network infrastructure, Case has to enter cyberspace in order to prevent his brain from being destroyed by an enemy corporate super-power. Uploading his body into the deadly computer network, he encounters an ultimate intelligence, Neuromancer, which, like Caire's *Cyberhead*, is essentially a complex human personality construct, an embodiment of the digital domain intertwined with human experience.

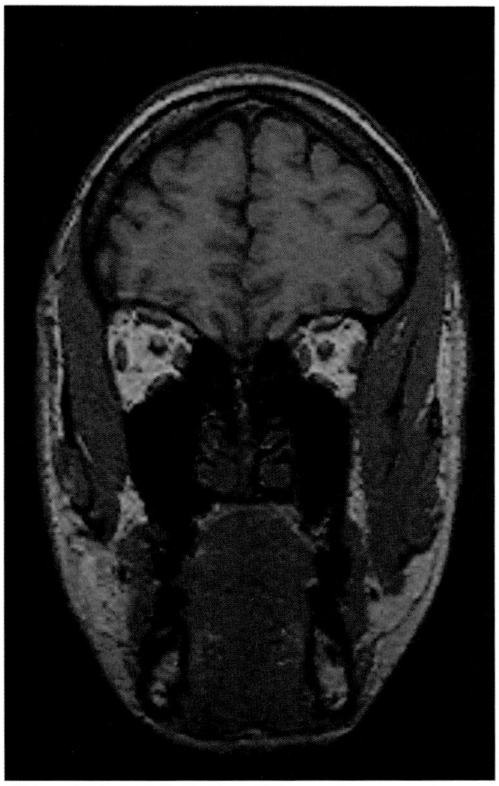

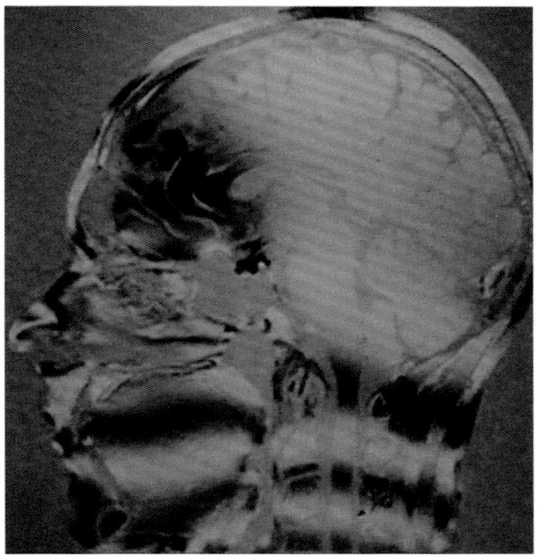

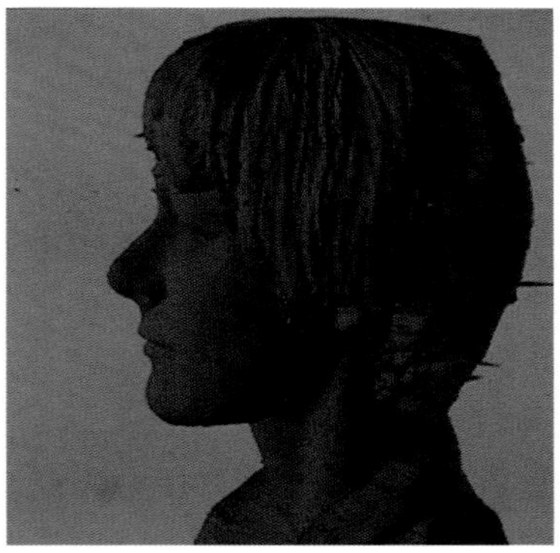

C.A.I.R.E.'s Cyberhead-Am-I-Really-Existing, *1994*

Cyberhead offers the viewer an opportunity to become part of cyberspace whilst maintaining an individual existence. As participants travel though a three-dimensional animation of a computer-reconstructed eye, optic nerve, visual cortex or nerve cell they experience a whole array of intellectual and emotional responses. *Cyberhead* is a site where the contradictory states of a player's existence may intertwine: the immortal and human, linear and multidimensional, singular and plural. In Caire's words:

My project is a fully immersed, interactive flight through a head reconstructed from Magnetic Resonance Imaging (MRI) data. This Virtual Reality journey runs in real time on a Silicon Graphics Onyx Reality Engine 2, with texture maps using a Fakespace BOOM 2C as a high-resolution stereoscopic viewing and navigating device. To build the world, the Sense 8 World Tool Kit software was used. 3D sound is generated by two Beachtrons from Crystal River.

Cyberhead was developed in the VR Laboratory and the AI Centre of SRI International in collaboration with the Lucas MRS Centre at Stamford University. Additional 3D CAD models and animation were created at Colossal Pictures: Spectrum HoloByte; and by Cyberware.

My primary goals were to create a rich, detailed virtual environment with convincing, high quality, real-time, reality-based visual images that were properly lit, smooth shaded, textured and anti-aliased. Directional sound was an equally important part of the work and experience. The human interface was designed to be simple, non intrusive and suitable for use by the general public. In relation to the audience, the goal was to create an entertaining experience that would make users think about such issues as how we interpret and associate the information we receive from our environment.

My motivation in doing this work was to explore new presentation paradigms made possible by technology. I had to address the problem of how to represent data that is not easy to represent: how to be immersed in, interact with and navigate through it and finally make the process aesthetically engaging and educational.

Caire is exposing her audiences to other possibilities of existence. She believes that our own biological systems change much more slowly than other systems so that we are anachronistically stuck with the natural limitations of our anatomical organisation. However, she does not view technology as a panacea. In cyberspace and VR, the infrastructure does not yet allow the artist to provide unlimited freedoms to the audience, only to provoke dilemmas and pose choices. Caire is redesigning these environments in terms that question our ability to relate to each other as human beings in society.
(Rachel Armstrong)

STELARC AND THE SUM OF HIS PARTS

WERNER HAMMERSTINGL

Wait! Is that the sound of falling solders? Is the king in trouble? Is he cornered? Turing's challenge is just getting interesting.[1] His human versus machine intelligence tests, and others inspired by them, will continue to attract contenders until the crown of chess is finally placed on a plastic box housing a fearsomely fast and cunning CPU (or more likely a whole colony of CPUs) and Kasparov will be just another statistic.

As the massively funded MIT labs work on yet another machine in the race to produce an artificial brain that will narrow the human-machine gap, we should not overlook Stelarc. When you look into his dark eyes you see the terminals of a brain that has already succeeded in viewing the AI versus human intelligence issue from a radically different perspective.

From his performances with industrial robots, to his 'involuntary muscle stimulation events', via the 'stomach sculptures', Stelarc's leitmotif has been the examination of the body. He has explored it from the outside, the inside, as hard body and virtual body, as data, as flesh, as eventscape, as a cybernetic device, as a site that locates philosophical, theosophical, political and scientific dimensions of interpretation and debate, and finally as one half of an interface whose other half is the machine.

He never appears to have wavered from this investigation, which he chooses to perform as an artist, not as a scientist. As such, he can work with a different, speculative rhythm. He is not weighed down by having to evolve completed theorems. He can slip in and out of a whole banquet of moral, technological, scientific, philosophical and medical dilemmas. And with each level of investigation we witness an increase in confidence, authority and maturity.

If at first Stelarc appeared to be something of a martyr for his cause, he has left that outdated model behind. He is presently authoring and orchestrating events that blur the distinction between the body's own intelligence and the body as 'object', controlled by communal intelligence: transforming himself into a meme within the huge operating system that constitutes the 'wired' population of the Internet. Mirroring the intentions of hundreds, if not thousands of individuals, and mediating the instructions transmitted by a colony of modems, Stelarc's body, an organic, and, 'in itself intelligent device', is publicly and visibly controlled by the invisible minds, desires and signals of all these 'wired puppeteers'. Mediation and translation take place in front of the audience in the site referred to as 'the body'.

The idea of surrendering one's real or perceived authority over one's body is foreign to some and downright blasphemous to others. This new terrain is likely to find more resistance than the earlier works, the spectacular, exotic nature of which appealed to the broader public. His new work and ideas, however, are veering dangerously close to an uncomfortable truth. Stelarc is empirically demonstrating that it is now physiologically and technologically possible to control a body. Like many visionary philosophical proposals Stelarc's will probably meet with the popular reluctance to embrace change.

I had never understood this 'believe the old, reject the new' attitude until I read K Eric Drexler's explanation of the principle on the basis of mental immune systems acting to prevent potentially damaging meme packages infecting our collective or singular brains.[2] The meme package of ideas that Stelarc presents is so fundamentally challenging to the status quo that he and his memes must be attacked to safeguard the very foundations of our present societal structure. If we eventually succumb to the idea that humanity can extend and share 'the body' in life, we must then naturally embrace speculations on what we are likely to do with, and about, the transition the body makes towards death.

Lately I've noticed myself getting interested in chess again, even playing the odd game against my computer. I often win but, then again, my Powermac is no 'Deep Blue'.

Notes

1 I refer here to the Church-Turing Hypothesis, which states that a machine can duplicate the functions of nature as well as mathematical ones. The 'Turing Test' is based on his proposal in 1950 that: 'if a computer could successfully impersonate a human being during a free-form exchange of text messages, then for all practical purposes, the machine should be considered intelligent'. See Charles Platt 'What does it mean to be human anyway?' in *Wired*, April 1995, pp132-37. Alan Turing also wrote the first chess algorithm for use in a calculating machine in 1948.

2 K Eric Drexler, *Engines of Change*, Fourth Estate (London), 1990, pp37-38.

FROM ABOVE: Remote: Event for Amplified Body, Laser Eyes, Automatic Arm, Third Hand and Video Shadow*;* Split Body, Scanning Robot Performance *(Photos: Tony Figallo)*

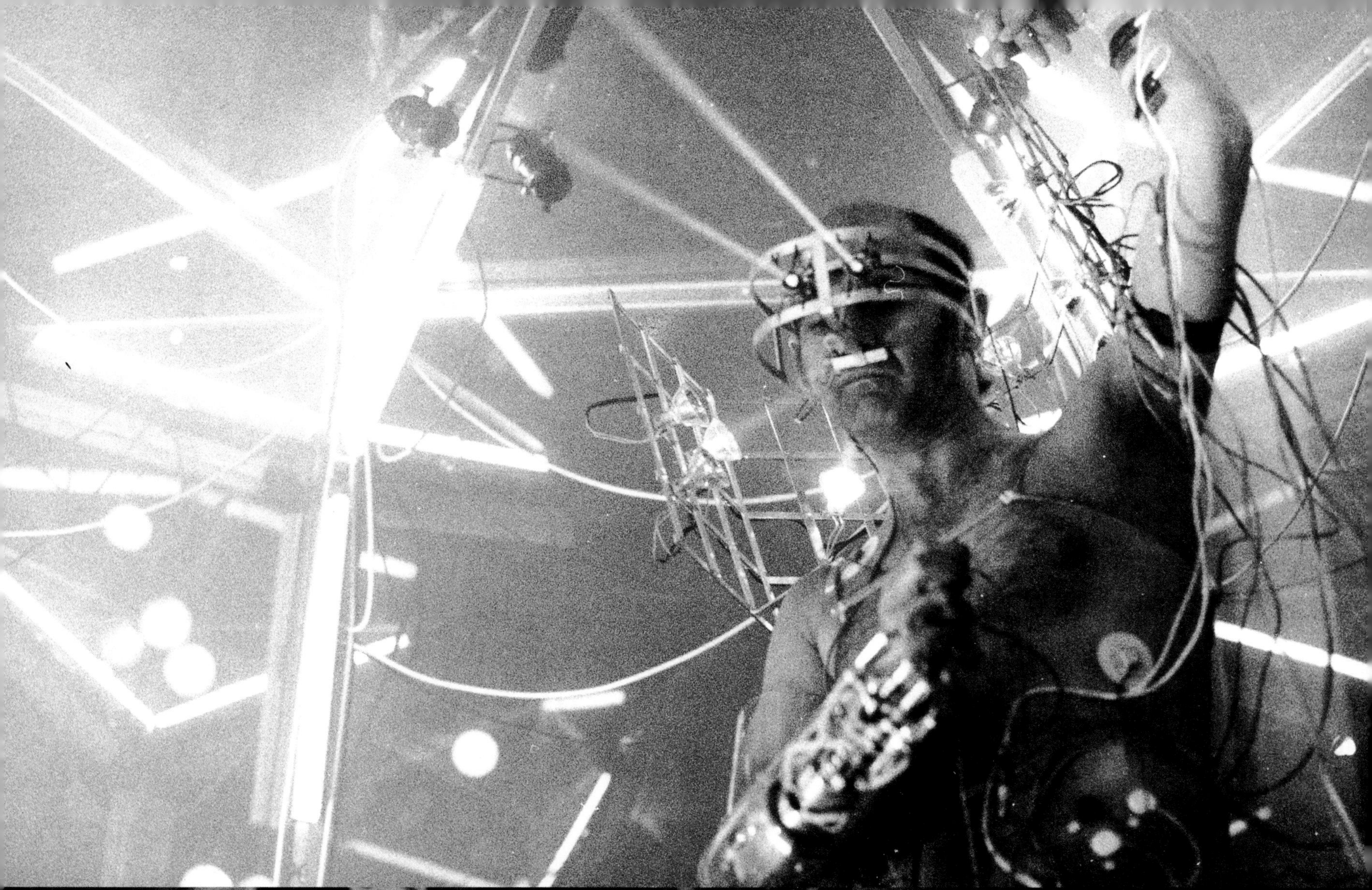

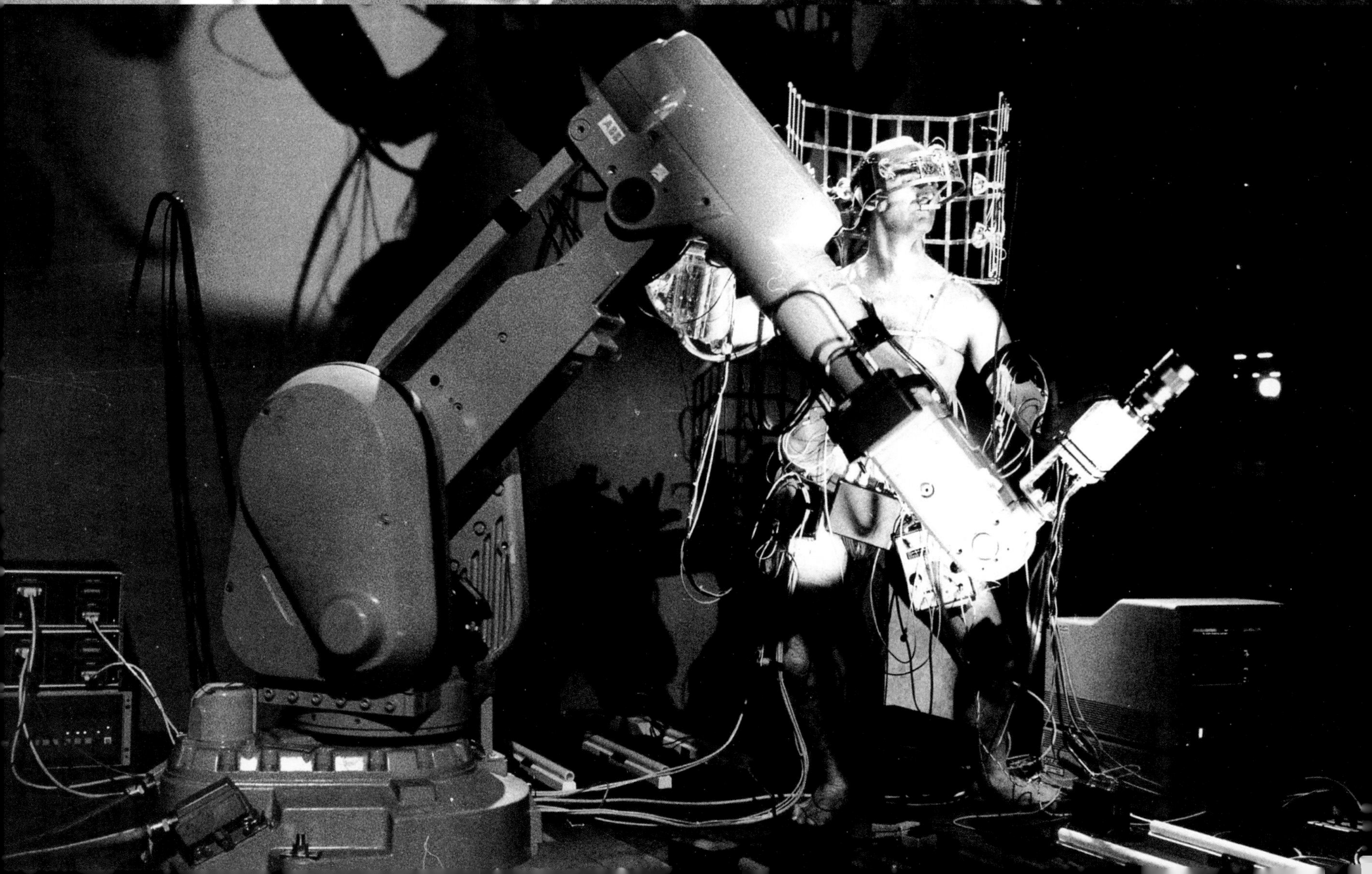

PARASITE VISIONS

Alternate, intimate and involuntary experiences

STELARC

Introduction

In previous actions the body has performed with technology attached (the *Third Hand* – actuated with EMG signals); technology inserted (the *Stomach Sculpture* – a self-illuminating, sound-emitting, opening/closing, extending and retracting mechanism operating in the stomach cavity); and Net-connected (the body becoming accessed and remotely activated by people in other places). The body has been augmented, invaded and now becomes a host – not only for technology, but also for remote agents.

Just as the Internet provides extensive and interactive ways of displaying, linking and retrieving information and images, it may now present unexpected ways of accessing, interfacing and uploading the body itself. And instead of offering a means of fulfilling out-moded, metaphysical desires of disembodiment, the Internet allows powerful individual and collective strategies for projecting body presence and body awareness. The Internet does not hasten the disappearance of the body and the dissolution of the self – rather, it generates new collective physical couplings and a telematic scaling of subjectivity. What becomes important is not merely the body's identity, but its connectivity – not its mobility or location, but its interface.

1 Fractal Flesh

Consider a body that can find its awareness and action in other bodies or bits of bodies in different places: an alternate operational entity that is spatially distributed but electronically connected. A movement that you initiate in Melbourne would be displaced and manifested in another body in Rotterdam; a shifting, sliding awareness that is neither 'all-here' in this body nor 'all-there' in those bodies. This is not about a fragmented body but a multiplicity of bodies and parts of bodies prompting and remotely guiding each other. Nor is it about master-slave control mechanisms, but feedback-loops of alternate awareness, agency and of split physiologies.

Imagine one side of your body being remotely guided whilst the other side could collaborate with local agency. You watch a part of your body move but you have neither initiated it nor are you contracting your muscles to produce it. Imagine the consequences and advantages of being a split body with voltage-in, inducing the behaviour of a remote agent and voltage-out, for your body to control peripheral devices. This would be a more complex and interesting body – not simply a single entity with one agency but one that would be a host for a multiplicity of remote and alien agents of different physiologies and in varying locations.

Certainly there may be justification, in some situations and for particular actions, to tele-operate a human arm rather than a robot manipulator – for if the task is to be performed in a non-hazardous location, then it might be an advantage to use a remote human arm attached to a mobile, intelligent body. Consider a task begun by a body in one place, completed by another body in another place; or the transmission of a skill. The body would not be a site of inscription but a medium for the manifestation of remote agents. This physically split body may have one arm gesturing involuntarily (remotely activated by an unknown agent), whilst the other is enhanced by an exoskeleton prosthesis to perform with exquisite skill and with extreme speed. This body would be capable of incorpo-

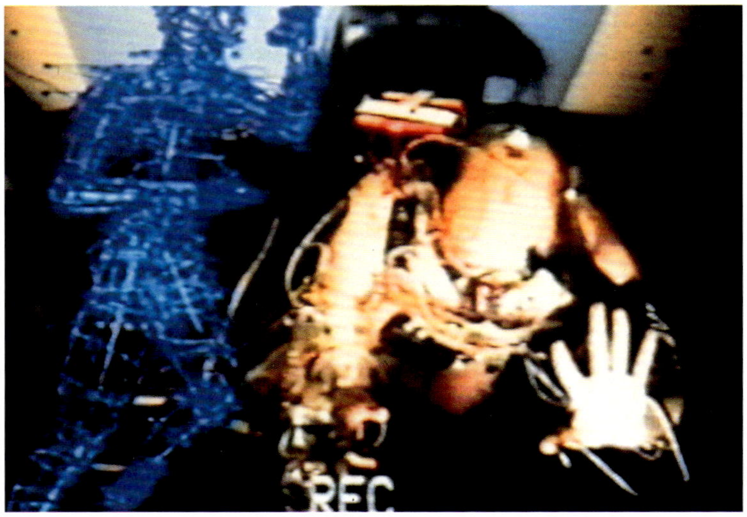

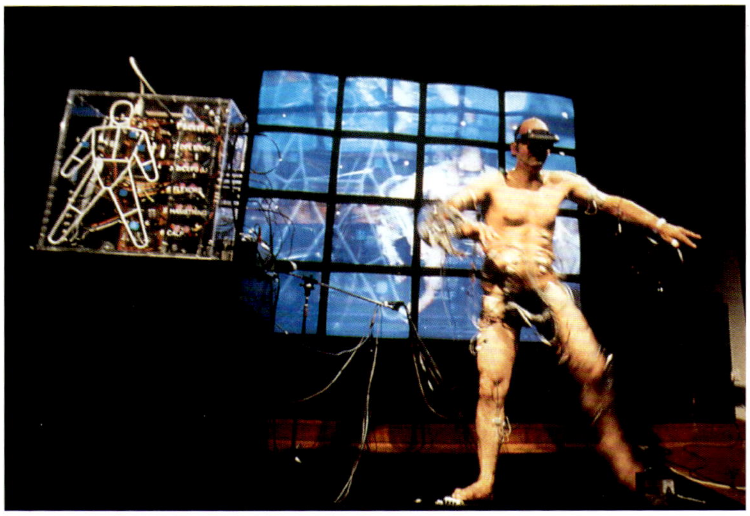

rating movement that from moment to moment would be a pure mechanical motion performed with neither memory nor desire.

2 Stimbod

What makes this possible is a touch-screen, muscle stimulation system. A method has been developed that enables the body's movements to be programmed by touching the muscle sites on the computer model. Orange flesh maps the possible stimulation sites whilst red flesh indicates the actuated muscles. The sequence of motions can be replayed continuously with its loop function. In addition to this choreography by pressing, it is also possible to paste sequences together from a library of gesture icons. The system allows stimulation of the programmed movement for analysis and evaluation before transmission to actuate the body. At a lower stimulation level it is a body-prompting system. At a higher stimulation level it is a body-actuation system.

Was it Wittgenstein who asked: if, in raising your arm you could remove the intention of raising it, what would remain? Ordinarily, you would associate intention with action (except, perhaps in an instinctual motion – or if you have a pathological condition like Parkinson's disease). With Stimbod, though, that intention would be transmitted from another body elsewhere. A two-way, tele-Stimbod system would create a possessed and possessing body simultaneously in the same physiology – a split physiology to collaborate and perform tasks remotely initiated and locally completed.

3 Extreme Absence and the Experience of the Alien

Such a Stimbod would be a hollow body, a host body for the projection and performance of remote agents. The body becomes hollow, erasing arbitrary distinctions between public, private and physiological spaces. Glove Anesthesia and Alien Hand are pathological conditions in which the patient experiences parts of their body as not there, as not their own, as not under their own control – an absence of physiology on the one hand and an absence of agency on the other. In a Stimbod not only would one possess a split physiology, one would also experience parts of oneself as automated, absent and alien.

In our Platonic, Cartesian and Freudian pasts this might have been considered pathological, and in our Foucaultian present we focus on inscription and control of the body. But in the terrain of cyber complexity that we now inhabit, the inadequacy and the obsolescence of the ego-agent-driven biological body cannot be more apparent. A transition from psycho-body to cybersystem becomes necessary to function effectively and intuitively in remote spaces, speeded-up situations and complex technological terrains.

During a Sexuality and Medicine seminar in Melbourne, Sandy Stone asked me what would be the cyber-sexual implications of the Stimbod system? Not having thought about it before I tried to explain what it might be like. If I was in Melbourne and Sandy was in New York, touching my chest would prompt her to caress her breast. Someone observing her would see it as an act of self-gratification, as a masturbatory act. She would know though that her hand was remotely and perhaps even divinely guided – ha, ha. Given tactile and force-feedback, I

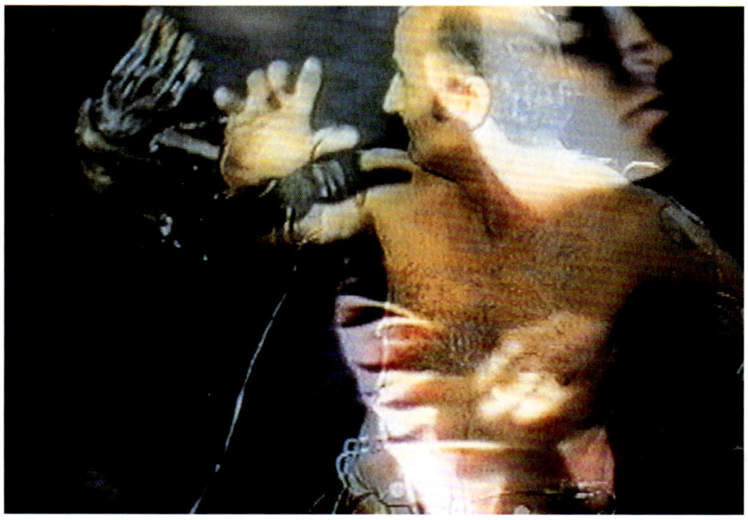

FROM L to R: Video still from Actuate/Rotate: Event for Virtual Body; Split Body: Voltage in/Voltage out *(photo: J Krpan); Video still from* Fractal Flesh

would feel my touch via another person from another place as a secondary and additional sensation. Or, by feeling my chest I could also feel her breast. An intimacy through interface, an intimacy without proximity. Remember that Stimbod is not merely a sensation of touch but an actuation system.

Can a body cope with experiences of extreme absence and alien action without becoming overcome by outmoded metaphysical fears and obsessions of individuality and free agency? A Stimbod would thus need to experience its actuality neither all-present-in this-body, nor all-present-in-that-body, but partly-here and projected-partly-there – an operational system of spatially distributed but electronically interfaced clusters of bodies ebbing and flowing in awareness, augmented by alternate and alien desires.

4 Ping Body / Proto-Parasite

In 1995 for Telepolis, visitors to the Pompidou Centre (Paris), the Media Lab (Helsinki) and the 'Doors of Perception' conference (Amsterdam), were able to remotely access and actuate the body in Luxembourg, using the touch-screen interfaced muscle stimulation system. ISDN Picturetel links allowed the body to see the face of the person who was moving it whilst the programmers could observe their remote choreography. Although people thought they were merely activating the body's limbs, they were inadvertently composing the sounds that were heard and the images of the body they were seeing – for the body had sensors, electrodes and transducers on its legs, arms and head that triggered sampled body signals and sounds and that also made the body a video switcher and mixer. And although people in other places were performing with the RHS of the body, it could respond by actuating its Third Hand, voltage-out from electrodes positioned on its abdominal and LHS leg muscles. This Split Body then, was manifesting a combination of involuntary, remotely guided, improvised and EMG-muscle initiated motor motions.

In *Ping Body – an Internet Actuated and Uploaded Performance* (performed first for 'Digital Aesthetics' in Sydney, but also for 'DEAF' in Rotterdam in 1996), instead of the body being prompted by other bodies in other places, Internet activity itself choreographs and composes the performance. Random pinging to over 30 global Internet domains produce values from 0-2000 milliseconds that are mapped to the deltoid, biceps, flexors, hamstring, and calf muscles – 0-60 volts initiating involuntary movements. The movements of the body are amplified, with a midi interface measuring position, proximity and bending angle of limbs. Activated by Internet data the body is uploaded as info and images to a Web site to be viewed by other people elsewhere. The body is telematically scaled-up to respond to the reverberating signals of an extended and external nervous system. The usual relationship with the Internet is flipped: instead of the Internet being constructed by the input from people, the Internet constructs the activity of one body. The body becomes a nexus for Internet activity, its activity a statistical construct of computer networks.

For the 'Virtual World Orchestra' in Glasgow the *Parasite* performance will involve Internet search engines that will scan, select and display bits of images and bits of sound randomly and continuously, providing a telematic optical and acoustical input for the body to augment its muscular proprioception. The body no longer sees and hears the real world but rather perceives the World Wide Web; a parasite body that moves and optically responds to an external and extended sensory spectrum.

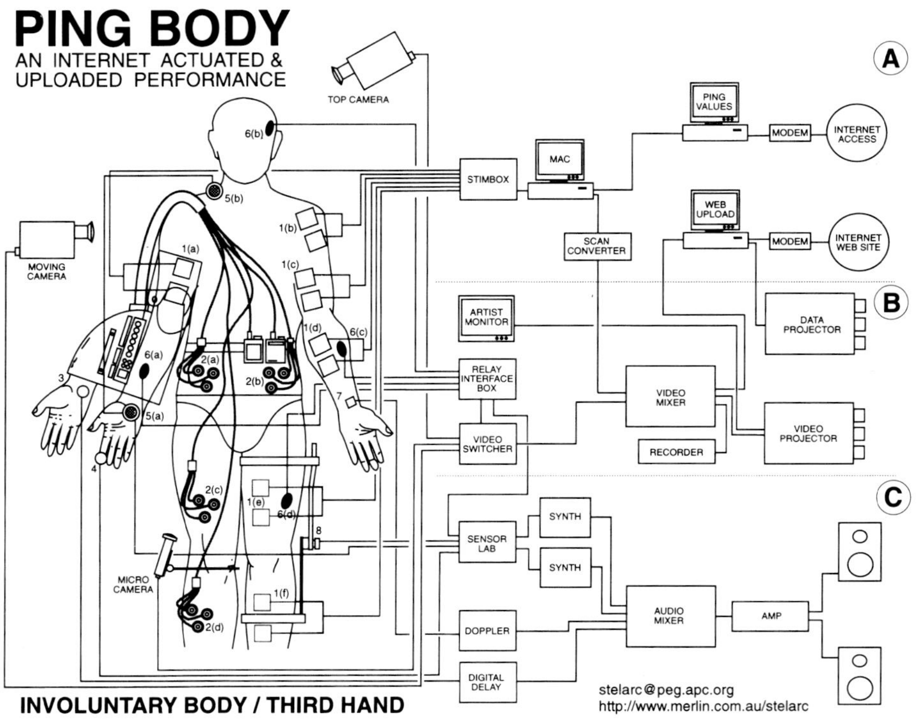

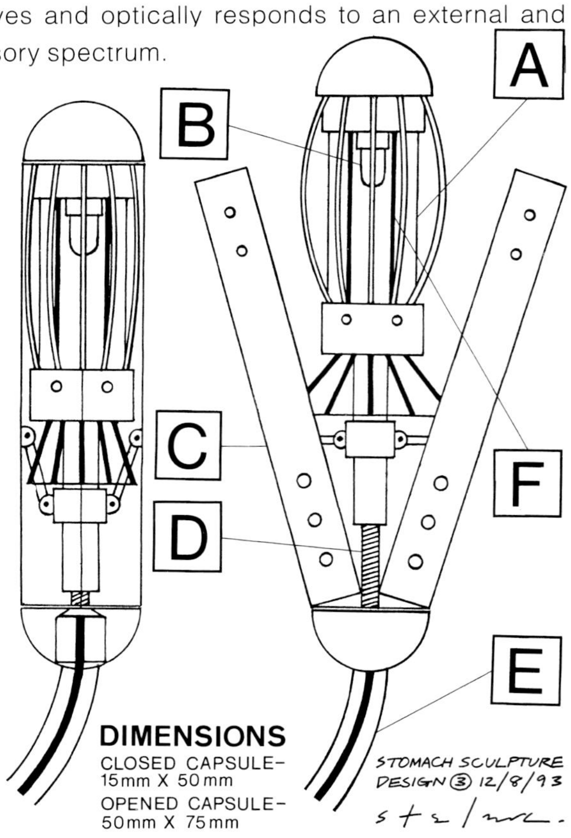

5 Phantom Bodies and Collective Strategies

Previously, connection and communication with other bodies on the Internet was only textual, with an acute absence of physical presence. This was not the experience of authentic, evolved absence that results in an effectively operating body in the real word – absence of the body on the Internet is an absence of inadequacy, that is, an inadequacy of appropriate feedback-loops. As we hard-wire more high-fidelity image, sound, tactile and force-feedback sensation between bodies, then we begin to generate powerful phantom bodies. The sensation of the remote body sucked onto your skin and nerve endings is palpable, collapsing the psychological and spatial distance between bodies on the Net. Just as the amputee experiences Phantom Limb, bodies will generate phantom partners, not because of a lack, but as an extending and enhancing addition to their physiology. It will only be through the construction of phantoms that the equivalent of our evolved absence will be experienced, as we function increasingly powerfully and with speed and intuition (a successful body operates automatically).

Bodies must now perform in techno terrains and data structures beyond the human scale where intention and action collapse into accelerated responses: bodies acting without expectation, producing movements without memory. Can a body act without emotion? Must a body continuously affirm its emotional, social and biological status quo? Or perhaps what is necessary is electronic erasure with new, intimate, internalised interfaces to allow for the design of a body with more adequate inputs and outputs for performance and awareness augmented by search engines.

Imagine a body remapped and reconfigured – not in genetic memory but rather in electronic circuitry. What of a body that is intimately interfaced to the World Wide Web and is stirred and startled by the distant whispers and remote promptings of other bodies in other places: a body that is informed by spiders, knowbots and phantoms. Consider the Internet structured so that it would scan, select and switch – automatically interfacing clusters of on-line bodies in real-time (the size and expertise of the clusters selected for the task to be performed). Can a body cope with the multiplicity of agents – a fluid and flowing awareness that dims and intensifies as they are connected and disconnected? Awareness and agency would be shifted and shared in an electronic space of distributed intelligence.

The Internet becomes not merely a means of information transmission, but a mode of transduction – affecting physical action between bodies: electronic space as a realm of action, rather than information. Consider a body whose awareness is extruded by surrogate robots in situations and spaces where no body could go. These machines with arrays of sensors, manipulators and hybrid locomotion would exponentially multiply the operational possibilities – scaling-up the subtlety, speed and complexity of human action. Perhaps what it means to be human involves not retaining our humanity.

Stimbod was developed with the assistance of Troy Innocent at Empire Ridge in Melbourne. *The Muscle Stimulation System* circuitry was designed by Bio-Electronics, Logitronics and Rainer Linz, Melbourne and the box fabricated with the assistance of Jason Patterson. *Tele-Stimbod*, *Ping Body* and *Parasite* software was developed by Gary Zebbington, Sam da Silva, Jeffrey Cook and Dimitri Aronov of the Merlin group in Sydney.

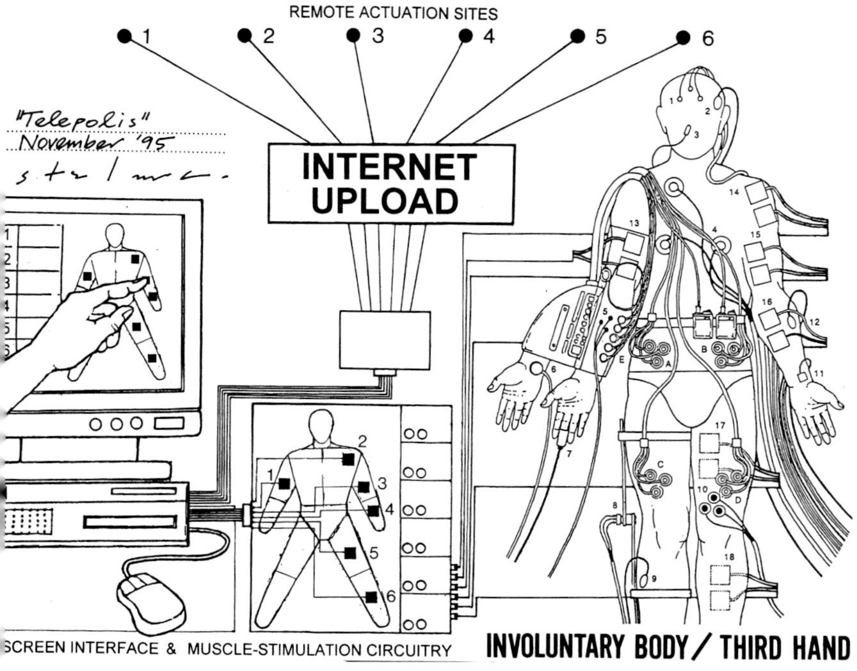

FROM L to R: Ping Body design; *Stomach Sculpture* design; Involuntary Body Third Hand designs

TAMAGOTCHI

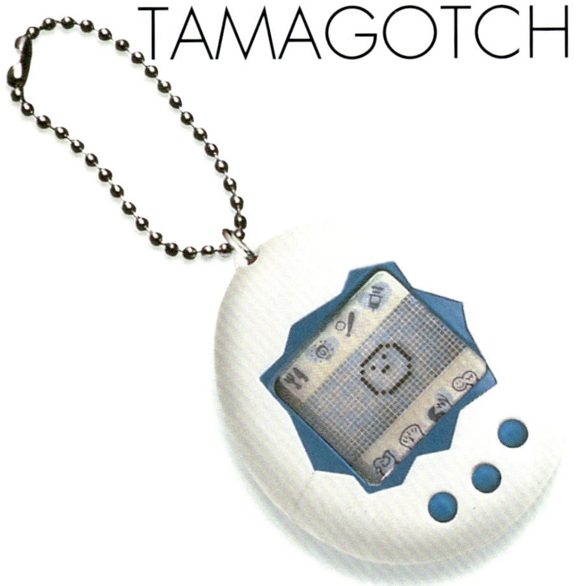

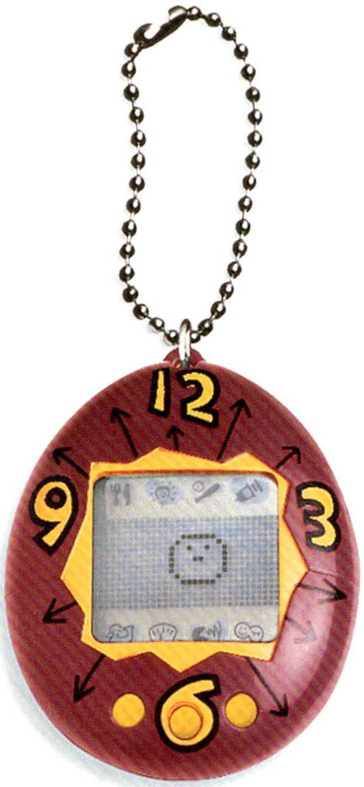

Just as medical scientists tinker with life's mechanisms in test tubes, computer scientists are attempting to create life on silicon-based, computer systems. Current intelligent automata, essentially composed of information, are given form by the computer hardware that serves as the vehicle for their emergence. The degree to which these artificial life forms resemble biological life varies. Many creators admit that they are merely simulations of different characteristics of existing creatures rather than brand new, self-organising systems.

The attribution of life-like characteristics in a number of systems has been so convincing that they have become commercially successful. Some artificial life forms are embedded in computer games, such as *Magic Carpet* by Bullfrog, where a whole community of nomadic people respond emotionally to the player's aggressive or passive presence.

The phenomenally successful invention, Tamagotchi, is an artificial life form modelled on the characteristics of a living, domestic organism. This Virtual Reality creature of Japanese origin hatches from its plastic egg, requiring human hand rearing by pushing buttons on the casing. After it hatches onto a tiny LCD screen, it demands a great deal of further human interaction such as feeding, cleaning, sleep and medical attention. With the appropriate input, the carer is rewarded by the development of the creature from baby to child and eventually to adulthood where it will exhibit one of seven possible character types in response to the way it is 'brought up'.

A properly cared for Tamagotchi may live for some time but many badly handled pets perish from neglect and abuse within a few days. The distributor, Bandai Co Ltd, reassure us that it does not matter if the first attempt at rearing Tamagotchi is not successful: we can always buy another. This is no longer strictly true: the demand for these devices is now so great that people from school-girls to business men are queuing overnight outside toyshops to obtain one. And although the retail price is about £10, they are auctioned on the Internet for up to £250.

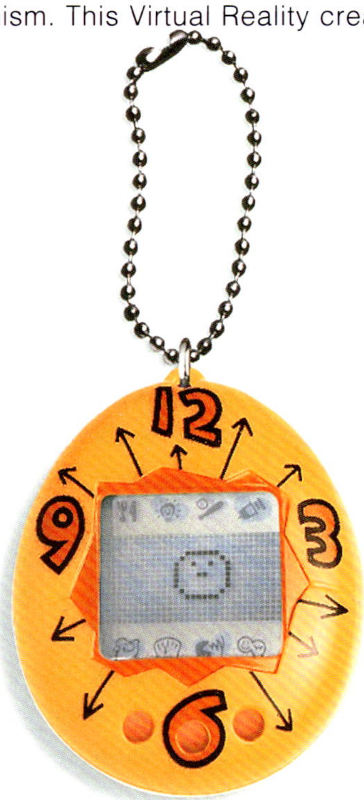

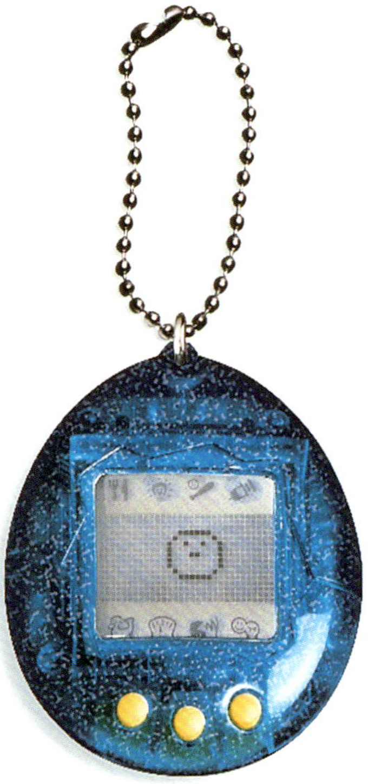

TINA KEANE

Transposition

This live work fuses several media elements to create a conceptually and physically engaging work. Video, sculpture, lighting, movement, sound and text are the components. The focal point is a sculptural piece of technology, the Travellator, which serves as the sole means of locomotion for a human screen of 25 men with naked backs.

This living screen provokes issues associated with the body: beauty, stereotyping, identity, notions of decay, death and the impersonal nature of science and industrial expansion.

The performers walk towards the Travellator in single file, step onto it and stand shoulder to shoulder with their backs towards the audience, never allowing a gap to appear. They allow the machine to transport them to the end of the walkway, where they alight and walk round to rejoin it with a hypnotic regularity. This repetition alludes to and reinforces notions of impersonal mechanical/technological presences in our lives.

Cinemascope images are projected onto the moving human screen throughout the performance. These relate to the idea of movement, both in the outer environment, and in the inner body. The impression created by the projected imagery is that of an X-ray, informing us of the histories of the mute performers and by extension, universal human histories.

Presented as a visually arresting installation-environment, the work offers a critique of the seductive and palliative effects of the alluring screen and its supporting ideologies.

X Ray
Nothing is static
X Ray
Interior and exterior
X Ray
Growth and decay
X Ray
Scientific inquiry
X Ray
Displacement of power

Equipment:
Two video projectors and playbacks
Black screen, approx 4.5 x 6m
Duration:
40 minutes

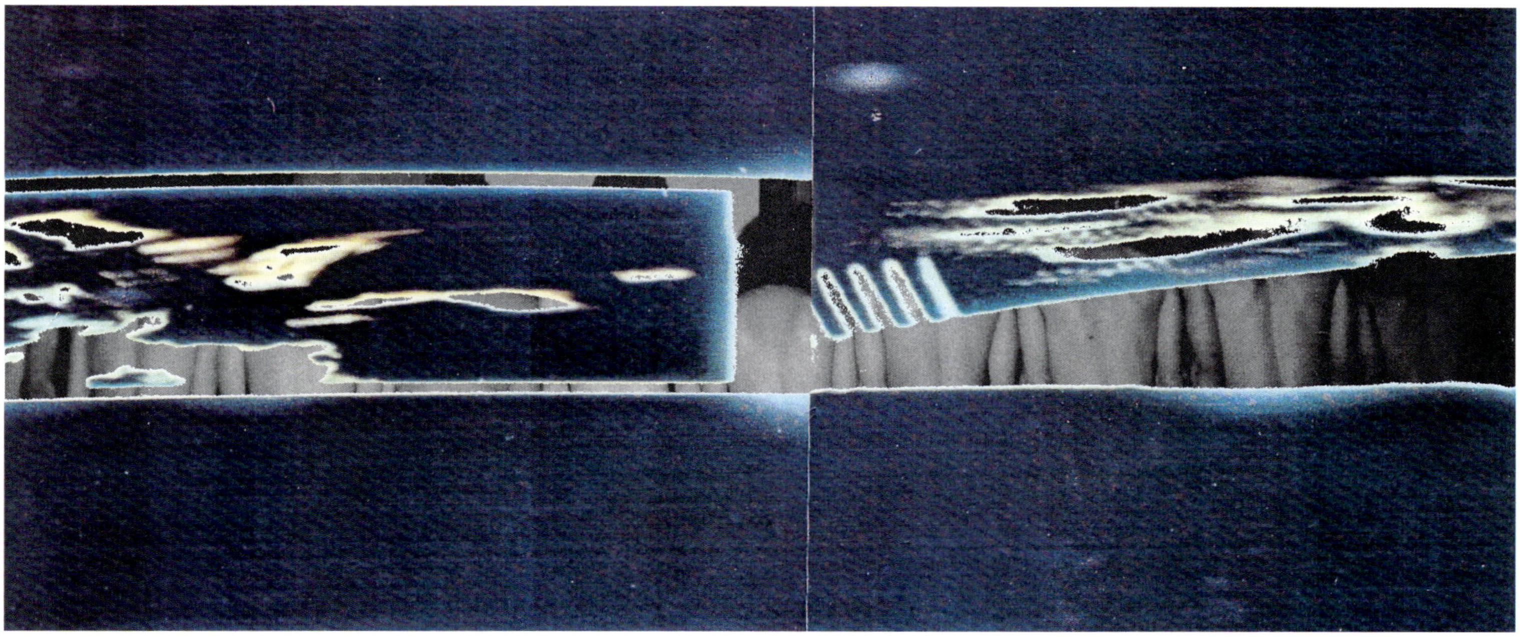

Transposition, May 1992

TAKING THE ALIEN EXAM

ROB LA FRENAIS

Some descriptions:

1 'In the 1950s ITT developed sensor technology that could literally display what a person was thinking. It was essentially a mind-reading-machine. It operated on the principle of picking up the electro-magnetic functions of human beings and translating those into an understandable form. It consisted of a chair in which a person would sit. Coils, which served as sensors, were placed around the chair – there were also three receivers, six channels and a Cray 1 computer, which would display what was on a person's mind, digitally on screen. A lot of hard work and a lot more computer crunching got things to the point where the computer could print out a dialogue . . . of what the person was thinking . . . The person would visualise something and a picture would actually appear on the computer monitor . . . which was used to decode the transmissions being generated from the person in the chair, [and] was then interfaced with an IBM 360 computer. This was, in turn, interfaced with the Montauk transmitter . . . A psychic would then sit in the chair in Southampton (Long Island) and relay via computer to the Montauk transmitter. This psychic would think thoughts, and the Cray 1 would decode them, they'd be put on a 32 bit radio link and sent to Montauk where they would go on to the IBM 360. The IBM computer would then broadcast it out of the transmitter and could build a thought form of what the psychic was thinking in Southampton. The devise was essentially a mind amplifier . . .

In his mind the psychic would concentrate on a solid object and it would appear somewhere on the base. Whatever he would visualise, the transmitter would transmit the lattice (or matrix) for, and build enough power to materialise whatever he was thinking of . . . In other words if he would hold an object in his hand and/or visualise it, it would appear at the given spot. They had actually discovered pure creation out of thought with the use of the transmitter.'

2 A group of sleepers lie half embedded in a shale-heap in the English countryside. They are attached to neurosynchronisers tuned to delta rhythms. They have inverted their sleep patterns so that day becomes night, night becomes day. An attendant occasionally sends signals to them via a transcutaneous electrical stimulator. They have learned codes in their waking moments so that there is a direct interface between the waking and the sleeping. It is transgressing a human boundary, like communicating with the dead. The TENS unit also stimulates awareness of lucid dreaming at moments of high REM activity. Later, they may set themselves special group activities, tasks to complete, destinations to visit, to be undertaken collectively in the 'dream' landscape. They are observed by the inhabitants of a town from a clock-tower through a telescope.

3 It is the middle of the night. Above a deserted mine shaft in a dark valley, four mediums are gathered in an open circle. They slowly, gently, enter, leave and re-enter trance states. They communicate with the dead. One of the mediums is wired up to a heartbeat monitor. The rhythm of the heartbeat reverberates around the mine shaft. The circle, dark, subdued, candle-lit, is surrounded by shadowy figures, some crouching, operating an array of high-technology equipment. The body warmth of the mediums is recorded by a heat-sensitive thermographic TV camera. There is high quality video and sound recording. A microwave transmitter sends the thermographic images across the valley where a group of witnesses and curious members of the public are gathered in a tent, situated by a floodlit, 16th-century decommissioned chapel. The mediums try to transmit their own images across the valley. One enters an intensive trance and his heartbeat, incredibly, speeds up to a continuous hum. All the recording equipment goes dead.

4 In a warehouse in a Yorkshire industrial town, a large white sphere is installed. Technicians in protective clothing slide individual participants, headfirst, lying on their backs on an oven-rack style apparatus, into the sphere. The hatch is closed and they are isolated, apparently floating in an intense light from above. After a while, the colour of the light subtly changes. Then there is a flash, almost inducing electrical shock in the subject. The light continues to flicker so brightly that it affects the retina whether or not the eyes are closed. Geometric patterns appear involuntarily: hexagons and octagons opening precisely as the computer programs, linked to the brain machine, vary the stroboscopic light. It is similar to the opening stages of an LSD experience. The patterns on the retina, the involuntary colour changes just outside the field of vision, the loss of control, are the same whoever the subject is. The experience is also similar to the many accounts of UFO abduction currently in circulation.

These are four accounts of radical ways in which humans and technology have interfaced. The last three are my descriptions of artworks by UK artists Bruce Gilchrist and Kathleen Rogers, and by James Turrell, the US light and earth artist. The first is an extract from *The Montauk Project – Experiments in Time.*[1] Apparently an account of a covert experiment taking place on a US military base during the 1960s and 1970s, it comes from the field of UFO and covert 'black budget' technology investigation.

Art and technology have become uniquely interfaced and the area of practice where these fields meet 'real' phenomena merits serious attention. Art provides a coherent belief system and structure that can be used positively as an alternative to science and religion (without attempting to replace or masquerade as either).

The three latter events are not scientific experiments, nor are they simply aesthetic performances. Rather, they could be seen as personal enquiries presented as artworks. They deal with areas that are in the realm of the spiritual but do not rely on established belief structures. They can be viewed by the public, but at least two thirds of the work is essentially hidden, taking place within the consciousness of the participants. The data collected is not scientific evidence as such, but is poetic debris that can be analysed and structured at will.

These four examples touch upon a number of debates around new attitudes to the human within the context of technology: the evolutionary consequences of the human/machine interface, the metaphor of the cyborg, the potentially transcendent nature of cyberspace and the iconic motif of the extraterrestrial (or extradimensional) manifestation. They inhabit all areas between science and mysticism, reality and fiction, artwork and thought experiment. Gilchrist employs art as a kind of vector in a bigger picture. Turrell explores the grey area between, on the one hand, the human subconscious and its resident demons, and on the other, nuts and bolts sightings and close encounters with apparently extraterrestrial phenomena. These events, in which process is more important than product, are all in a sense, unfinished, open-ended works.

Whether the secret experiments of the first example took place or not is less important than the poetics of the imagery (though the research community, to whom these are matters of life and death, would probably not agree). They emerge from a nexus of post-Cold-War, conspiracy-theory ufology, and a form of technological desire specific to male Americans of a certain age, usually with military backgrounds. They satisfy theories of 'reverse engineering', which claim that secret, unelected governments have had access to alien technologies and are slowly integrating them into terrestrial applications such as the Stealth bomber and the top-secret 'Aurora' military plane. They also bear an indirect but uncanny resemblance to the 'Ganzfeld' techniques used by the Department of Parapsychology at Edinburgh University, which enhance and verify the testing of telepathic abilities.

The works by Bruce Gilchrist and Kathleen Rogers are from the EarthWire project, curated by Tracey Warr and myself, which took place in a remote rural location in North East England in 1994. Gilchrist's sleep project was a result of a residency by the artist at the sleep research laboratories of the University of Austin, Texas. The intended experiments with lucid dreaming were hampered by the reversal of sleep patterns, which produced extended altered states during their waking hours, including a hypnagogic condition that left them convinced they had not slept. Gilchrist will continue this work with an experiment/installation using a bioactivity translator linked to a brain machine designed to observe changes in the artist's skin voltage during various states of consciousness.

Kathleen Rogers is an electronic artist who has worked with VR applications and parapsychology. Her collaboration with mediums and employment of the thermographic camera and the microwave transmitter follow a tradition of using technology to 'test' phenomena. However, the discourse between the two, though compelling, can often be unstable, since the reliability of the 'evidence' produced seems to be affected by the presence of powerful phenomena.

James Turrell's *Gasworks* installation, presented by the Henry Moore Sculpture Studio at Dean Clough, Halifax, provides an indication of where this kind of work might be leading on a wider scale. Famously, Turrell is constructing a vast earthwork at the Roden Crater in Arizona for the new millennium. Involving the ancient technique of 'celestial vaulting' on a massive scale, the work resembles nothing less than a landing site for UFOs. Unusually for an internationally reputed artist who is almost an establishment figure, Turrell openly admits to more than a passing interest in the UFO as both metaphor and scientific possibility. Like its predecessor *Alien Exam*, *Gasworks*, both in its direct engagement with the technology of human consciousness and in its resemblance to an abduction scenario, is quite openly linked to this interest.

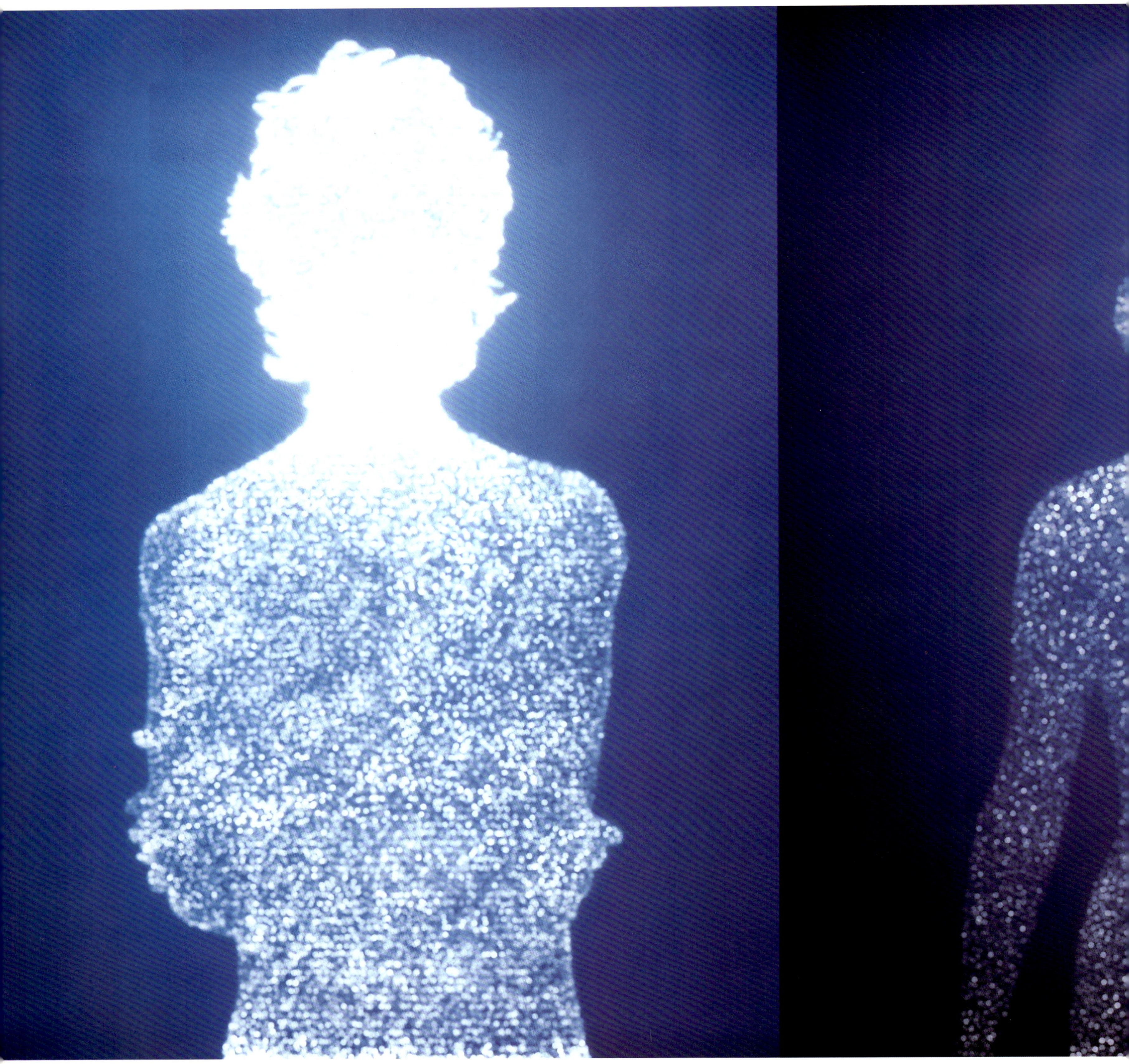

In *UFO's and the Alien Presence*,[2] which discusses the possibility of government cover-ups of UFO incidents, the political researcher and ufological writer Michael Lindemann has referred to the 'conspiracy' of active ridicule concerning UFO-related matters. The same could equally apply to parapsycological matters, and indeed anything involving the extra-scientific investigation of human consciousness. With an alternative belief system to that of science and religion it may be that artists can be immune to this conspiracy. Certainly, when those with the stature and resources of Turrell fly in the face of the culture of ridicule, artists wishing to continue their personal enquiries into such areas are provided with a powerful ally.

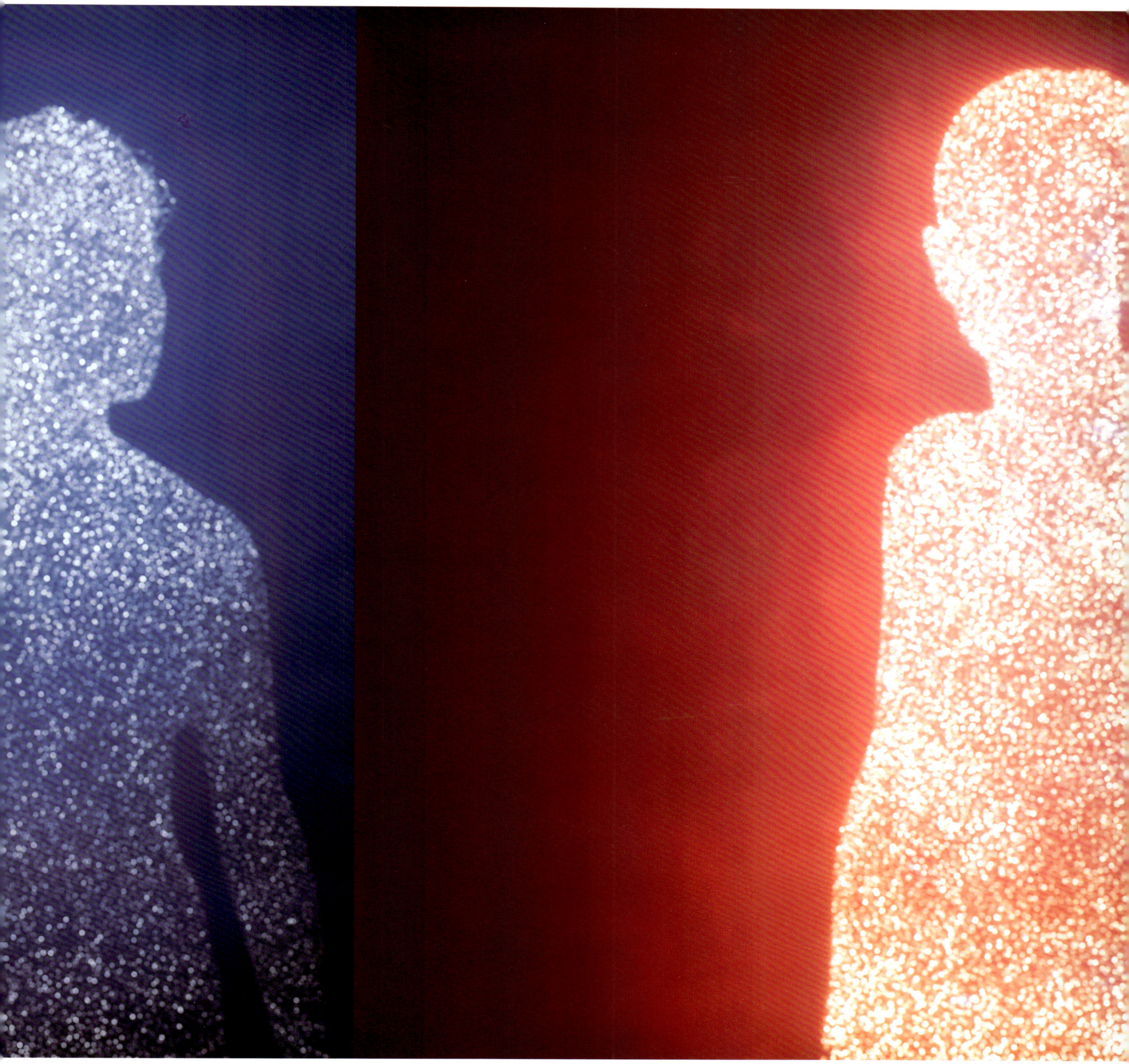

ABOVE: *Christopher Bucklow, Images from the series* Guest, *1993-95, unique Cibachrome photos*

Notes

1 Preston B Nichols with Peter Moon, *The Montauk Projects – Experiments in Time*, Sky Books, (New York), 1992

2 Michael Lindemann, *UFOs and the Alien Presence – Time for the Truth,* The 2020 Group, Santa Barbara, California, 1990.

MARIKO MORI

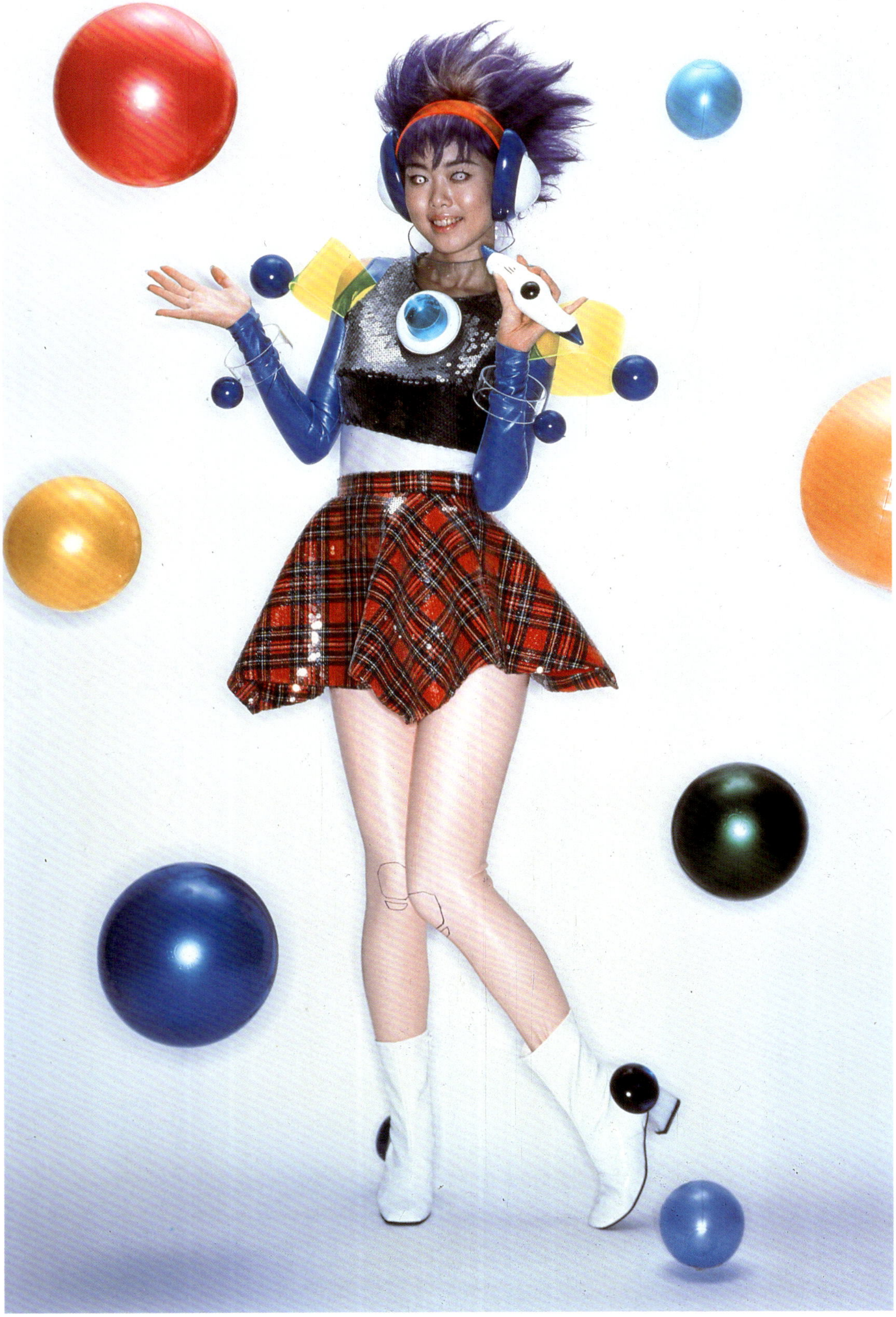

Mariko Mori uses computer techniques, along with holography to experiment with portraiture. Her photographs suggest celestial bodies composed of a substance beyond the gravitational laws of matter and promise a possible immortality and transcendental image in virtuality. LEFT: Birth of a Star, *1995, 3D Duratrans print, acrylic, fluorescent lights, 180 x 120cm; OPPOSITE, FROM ABOVE:* Entropy of Love, *1996, Cibachrome print, aluminium, plywood, formica frame, 5 panels: 3m x 6.5m x 8cm;* Last Departure, *1996, Cibachrome print, aluminium wood, smoke aluminium, 2m x 3.5m x 8cm*

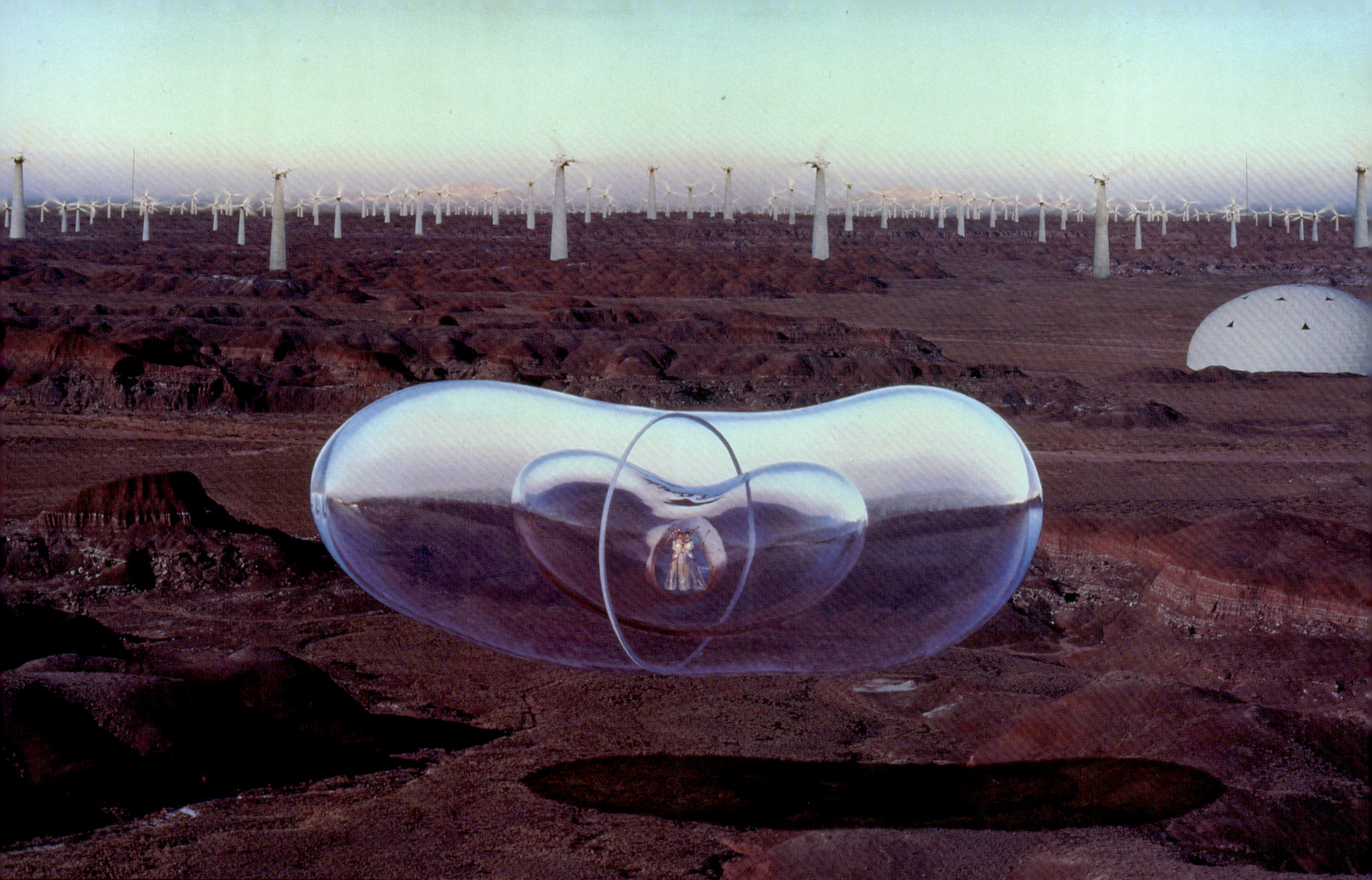

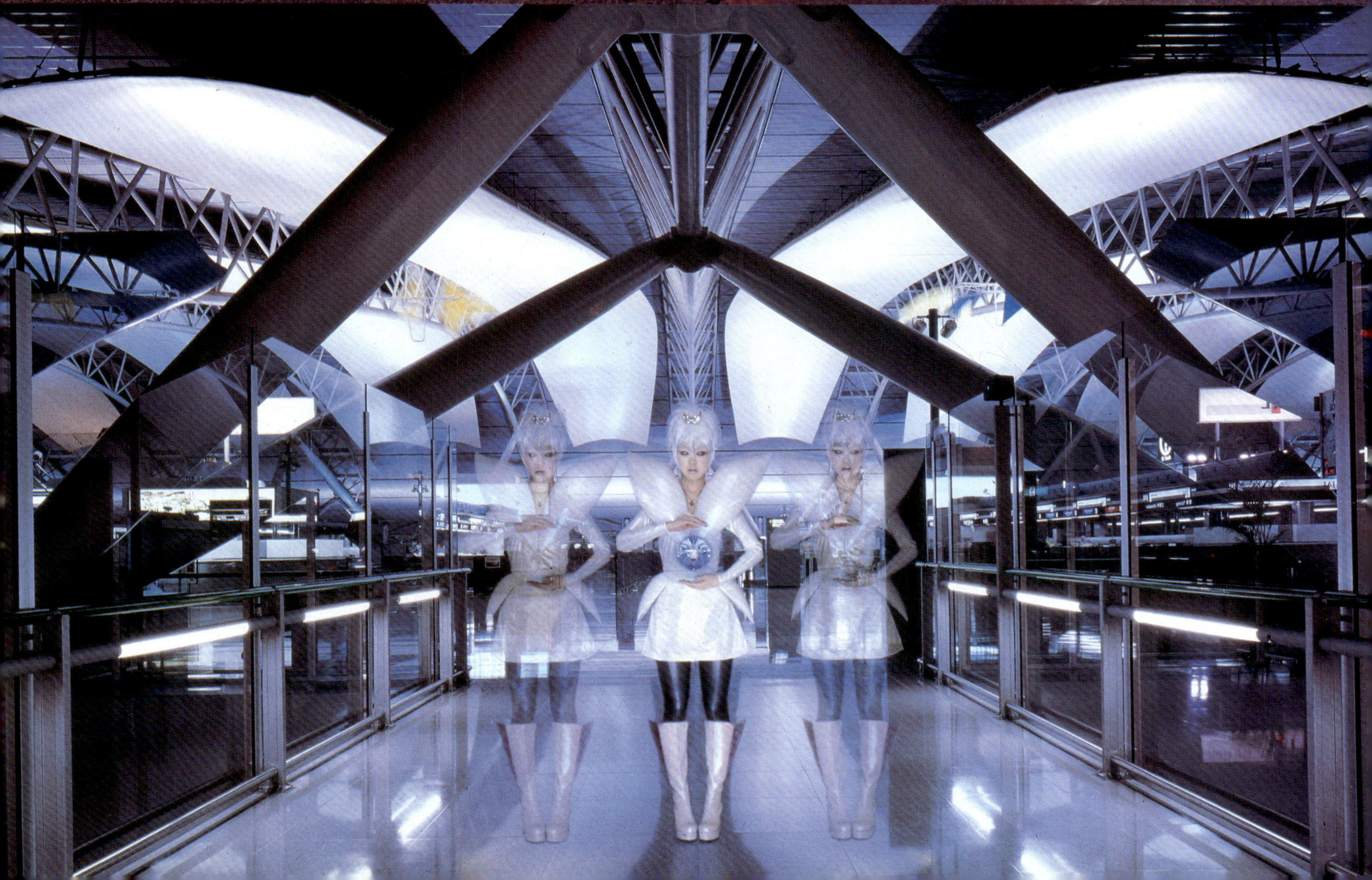

DAVID BOWIE

Alien Archetypes and Proteans

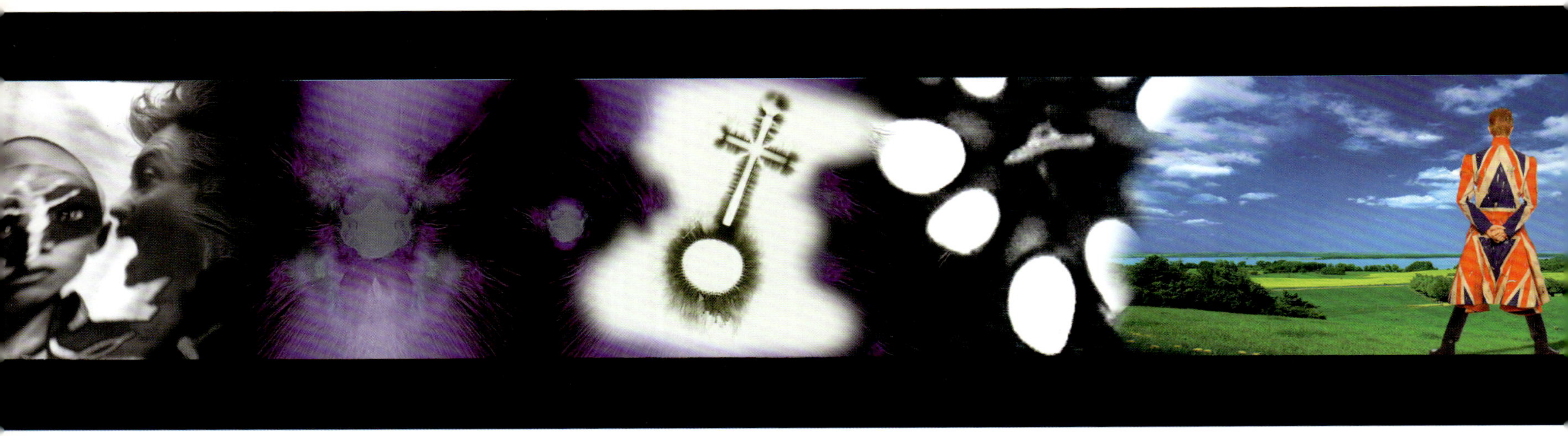

It is possible that the alien archetype is the subconscious projection of a perfect celestial body emanating from the human collective conscious. We may wish to mimic this new ideal by undergoing surgical operations to enlarge our heads, thin down the muscles in our bodies, remove our genitals, and change our eyes to black. Following this transformation, the physical symbols of our metamorphosis will convey that we are of a higher order of life than the old, unmodified humans – that we now possess enhanced intellectual capabilities such as psychic powers, are part of a technologically sophisticated culture, and have successfully abandoned 'redundant' behaviours like emotion.

However, there is an alternative to this fixed ideology of the alien, which has been extensively explored in the work of David Bowie who offers a continuously metamorphosing, protean body as his radical vision of extra-terrestrial life.

Ever since 'Space Oddity' in 1969, Bowie has been fascinated by alien cultures and their possible relationship to humanity. This album coincided with the Apollo 11 mission to the moon and was followed by 'The Rise and Fall of Ziggy Stardust and the Spiders from Mars' in 1972, whose central character was an alien from outer space perceived by earthlings as a rock and roll star. In many of Bowie's subsequent albums – 'Diamond Dogs', 'Scary Monsters', 'Tonight' and 'Day In Day Out' – there are references to alien experiences and extra terrestrial encounters and visions of future beings.

The music and images for the 1995 album and tour, 'Outside', were conceived in collaboration with the graphic artist Davide de Angelis and the musician Brian Eno. Bowie's starting point had been 'The Diary of Nathan Adler', an essay that he had originally written for *Q* Magazine, and which is reproduced with photographs and artwork in the album booklet. In this story, Detective Professor Nathan Adler of Art Crime Inc is an intelligence agent living in the year 1999. He is investigating the murder of a 14-year-old victim, 'Baby Grace', as one in a series of 'art crimes', which have been endorsed by collectors and funding bodies as acceptable practice, inseparable from other forms of artistic expression. Throughout the diaries, the detective questions the motives of the crime and interviews a number of disturbing suspects. All the characters are played by the protean David Bowie in a variety of guises, and references are made to the actual practices of contemporary body-performance artists such as Ron Athey.

Bowie's examination of our perceptions of a global human monoculture and those who deviate from its culturally established criteria, has been explored in an alternative way in his new album, 'Earthling' (1997). This collection, a melt-down of music types, writing techniques and fluid images, takes rock music into the future.

Bowie's shifting images and impressions are suggestive of a morphogenic process in society where instability and culture clashes are necessary for evolutionary change, encapsulating the aesthetic paranoia raised by the prospect of trans- and post-humanism.

In the following article, Davide de Angelis describes how he helped Bowie achieve his protean image both on 'Outside' and 'Earthling'. *(Rachel Armstrong)*

DIGITAL FRANKENSTEIN

Collaborations with David Bowie

DAVIDE DE ANGELIS

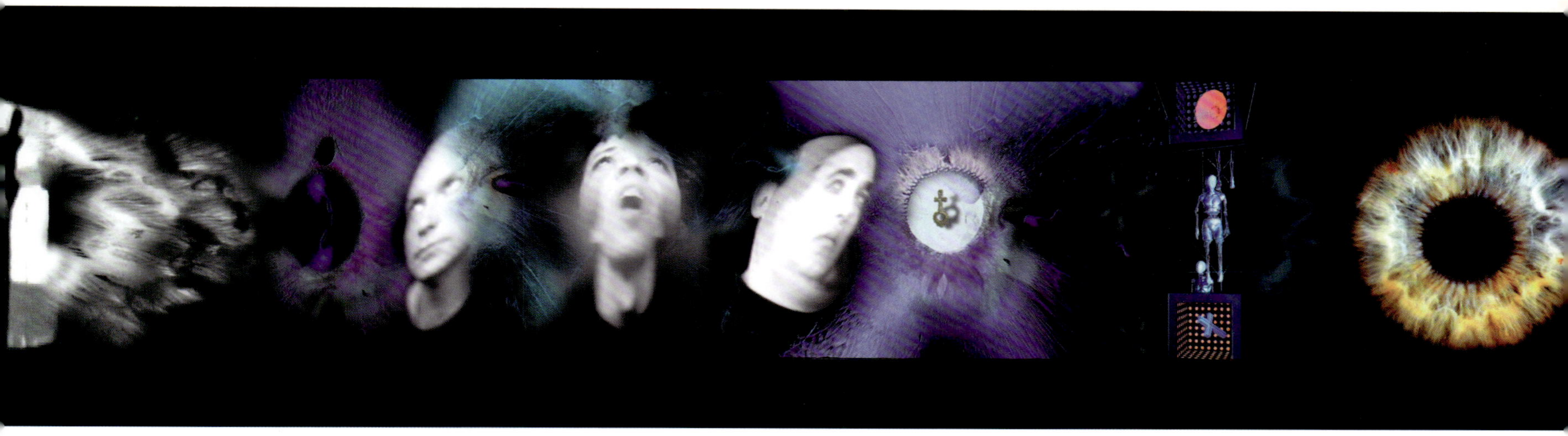

Outside

My collaboration with David Bowie began when he called me to say that he had seen an article containing my work and liked the way I was expressing ideas through the visual medium.

I had been experimenting with the use of randomisation in images, language, typography and symbolism, in order to create something with meaning from the many possibilities. The objective I had always been working towards was to create a feeling of spontaneity through the 'cold' machine. At this time I was filling redundant computers with the kind of gremlins most corporations spend millions trying to destroy.

In what turned out to be a long conversation, Bowie and I came to the conclusion that there were several similarities between the way he was using and deconstructing music and lyrics, and the methods I was exploring on the visual side. We decided to meet up and continue the train of thought.

Bowie came to visit me in my studio, bringing with him 'The Diary of Nathan Adler' and several photocopies of the work of outsider artists. In this first meeting we began to construct a rough outline of how we could produce imagery together.

Through our discussions it became clear that whatever we created would have to be a multimedia project in the true sense. We were aiming to achieve a body of work with the capability to be translated across a host of different media: print, CD-ROM, CD-PLUS, Internet, stage, video and TV.

Bowie's story centred around a group of characters and was subtitled, 'A non-linear Gothic Drama Hyper-cycle'. Following some early experiments, we decided to use Bowie himself in the role of each character. In a second meeting we added to the team photographer John Scarisbrick and stylist Jennifer Elster. Scarisbrick had mainly worked for fashion magazines, producing photographs in which his models are distorted with extreme fisheye lenses. This, we could use to help add a strange sense of unreality to the characters.

The final imaging involved a number of different procedures ranging from hand-drawn elements, photography and morphing techniques. I had sketched out rough ideas before the photographic shoot, which took place over three long days in London, and I used these as a starting point. This was perhaps the most nerve-racking part of the whole project: not only was I art directing the shoot, but we were also being filmed by a documentary crew. I had to ensure that we stuck to a strict timetable, producing one or two test shots before the final take: re-shooting was not an option.

One of the most interesting aspects of creating the final images was the use of the Internet. To begin with I would work up a rough composition on the screen and transmit it to Bowie in New York. Using a conference line we would discuss the various aspects of each image. He would sometimes scribble instructions on an image and flow it back down the line to me. This often continued over a number of hours until we were either satisfied with how things were going or too tired to make rational decisions. So I effectively became a kind of 'digital Frankenstein', piecing the characters together with a host of transplanted body parts from a range of donors both real and computer generated.

Earthling

The 'Earthling' project began when I received a package from Bowie containing a series of images and a sketch of an idea

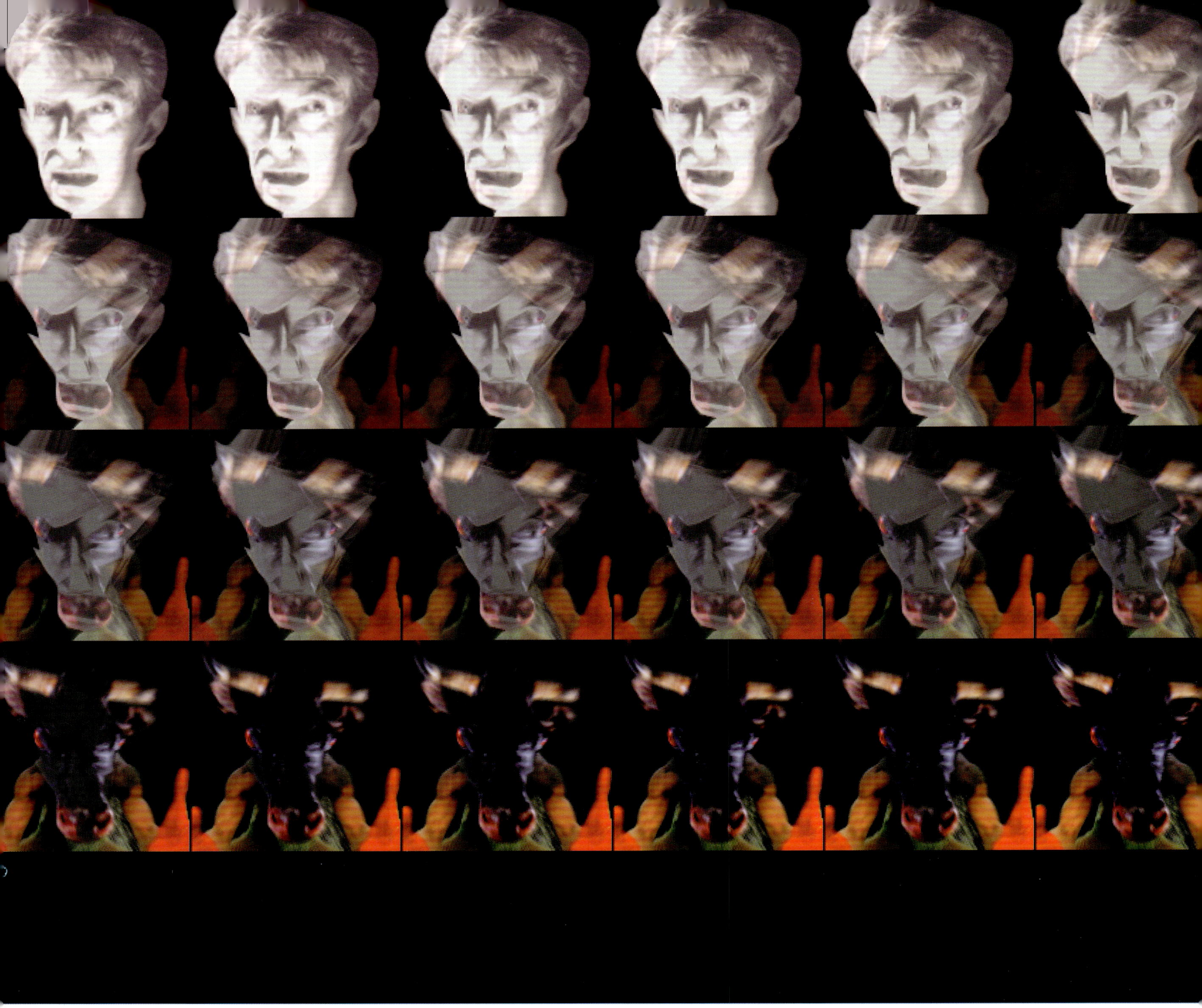

on watercolour paper. The package also included several black and white photographs taken by Frank Ockenfels of Bowie and the band members. The quality of image was more in keeping with a Caravaggio painting than something shot using a high-tech lens, but displayed the anti-gravity space glow of early Apollo mission images. We both picked up on this strange mixture.

Eventually we began to compare the compositional structure of monastic scripture and computerised digital display units. This gave rise to the typography treatment. The goal was to emphasise how similarities can occur between the patchwork-like structure of illuminated scripts, where texts are often a jumble of additions and afterthoughts, and the almost illegible, fast-flashing information on digital displays, which again, are often travelled through in a non-linear manner.

The artwork presents a continuous strip of imagery with different components blending into one another. The cross, with its string-like energy forces revealed through Kirlian photography, blends into a photo of a flying saucer. This underlines the energy created through our belief structures and the way

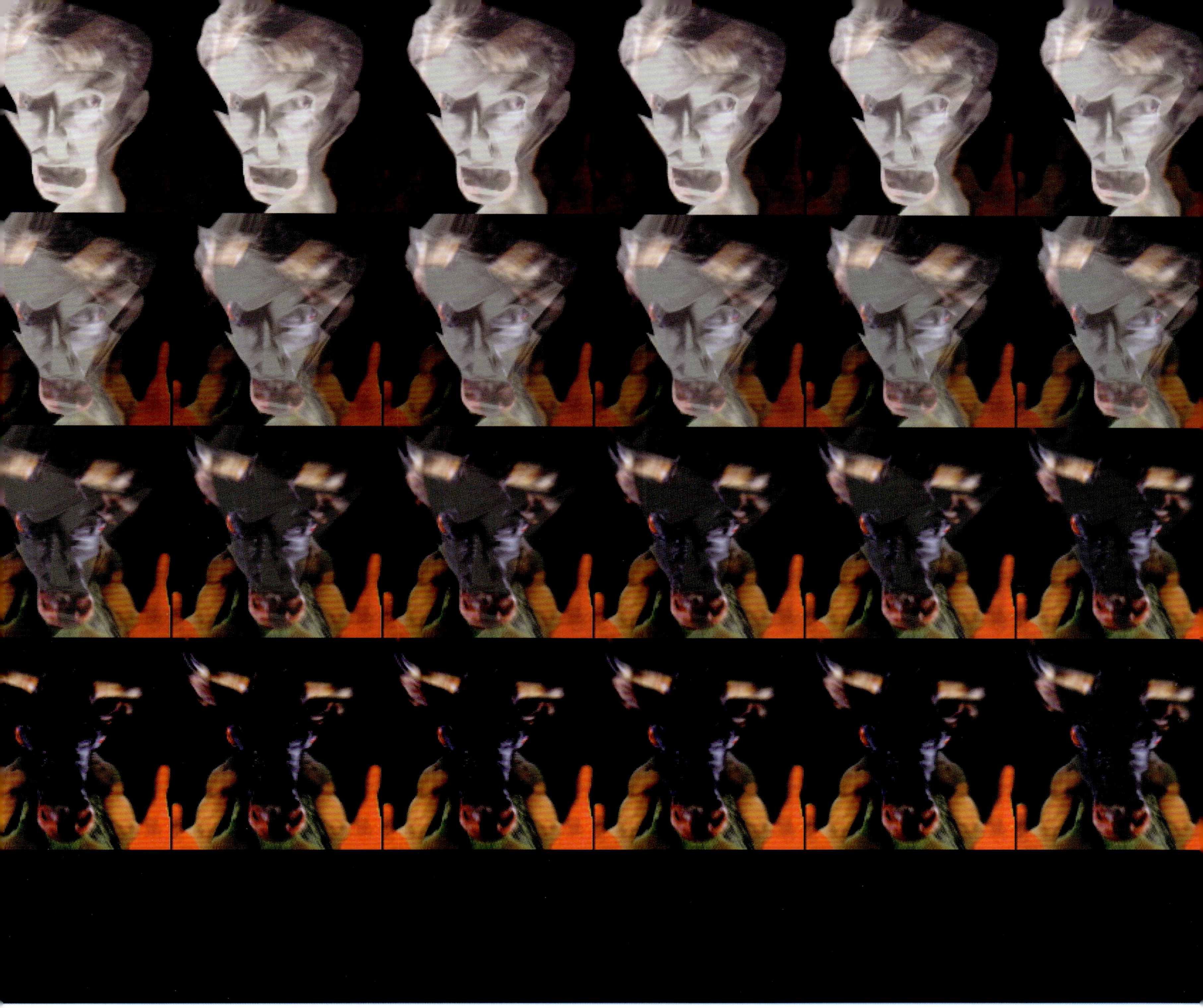

in which the alien has infiltrated our cultural vision.

After composing all of the images that ended up in the final artwork, I needed something to bind them together whilst creating the feeling that without this binding they would all fly randomly apart, disappearing into infinity without trace. Using strands taken from a Kirlian macro shot of Bowie's eye, I created on the computer a kind of cosmic/ectoplasmic web to link the images together.

The use of the familiar arranged in such a way that it becomes unfamiliar is a clear feature of the artwork. This is particularly evident on the cover where Bowie, standing with his back to the camera and wearing the Union Jack, is superimposed against the English countryside. This image has been colour manipulated, giving it the appearance of a huge, artificially lit eco dome in a distant galaxy. (Further information can be found at: www.davidbowie.com and at www.joycatcher.com.)

pp78-81: Earthling, *art work, 1997; OVERLEAF:* Little Wonder, *1997, video stills, director Floria Sigismondi, sculptural collaborations with Tony Oursler*

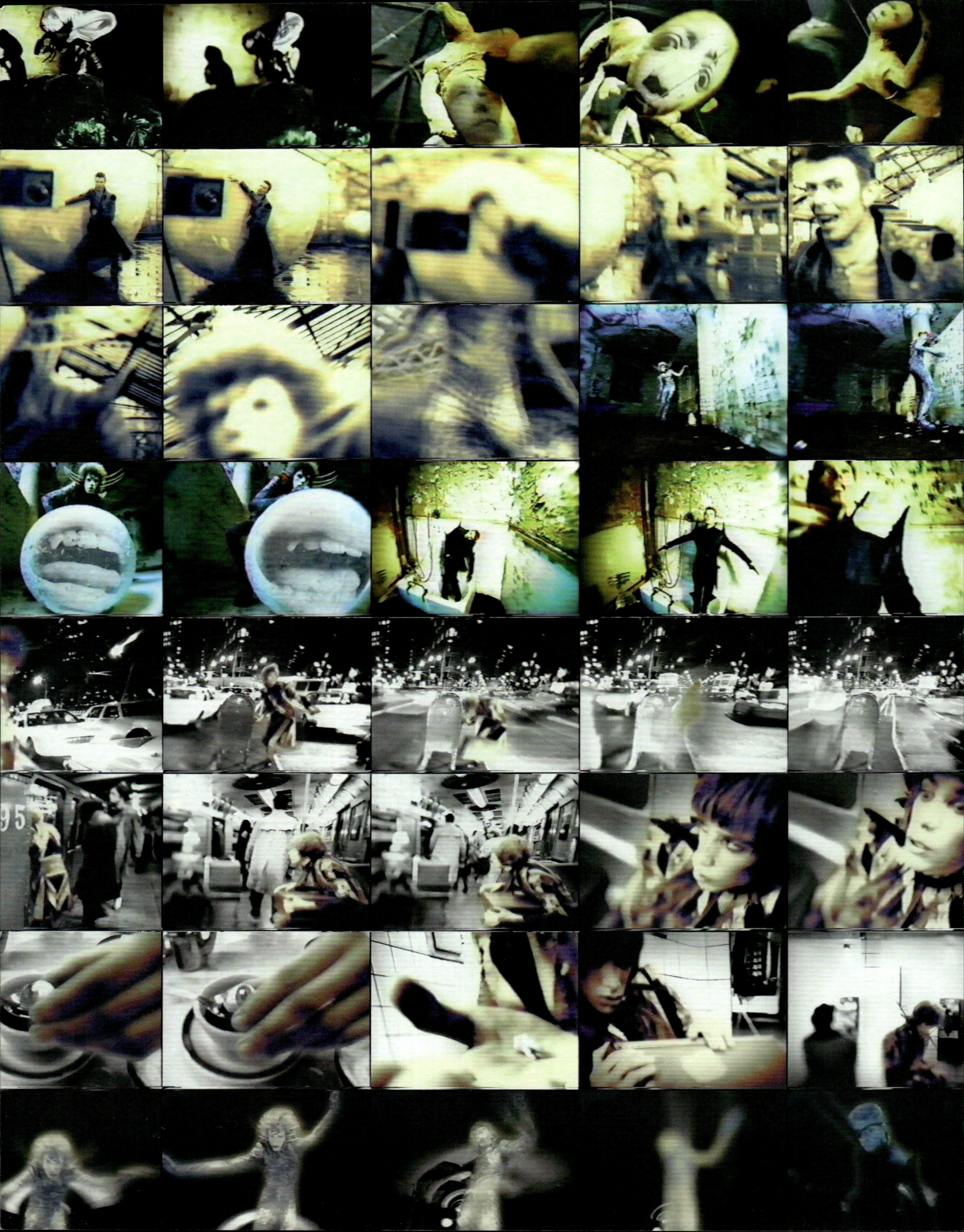
95

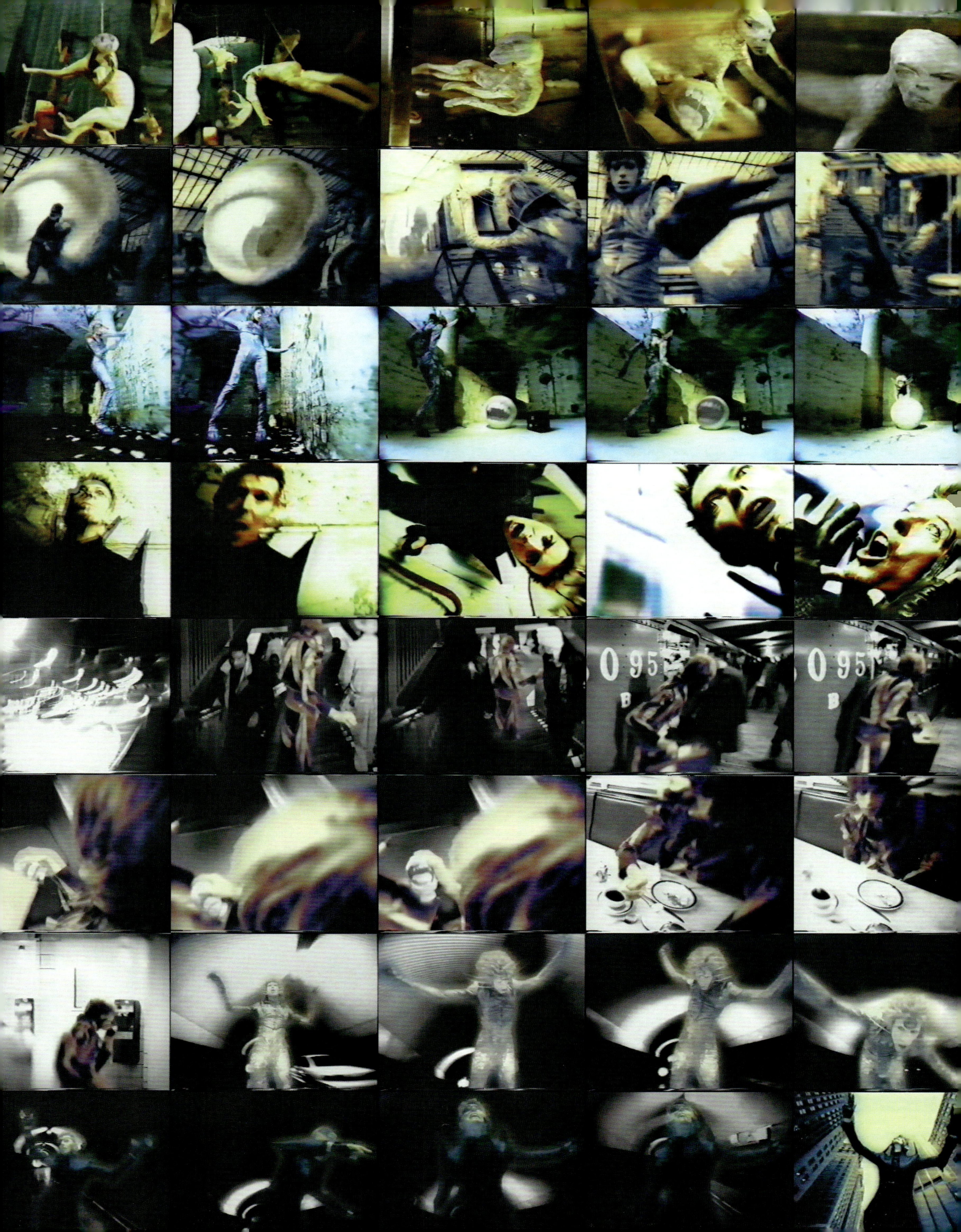

Michael Buhler, FROM L TO R: Visitors, *oil on canvas, 117 x 68.5cm;*
The Subject, *oil on canvas, 83.8 x 55.8cm*

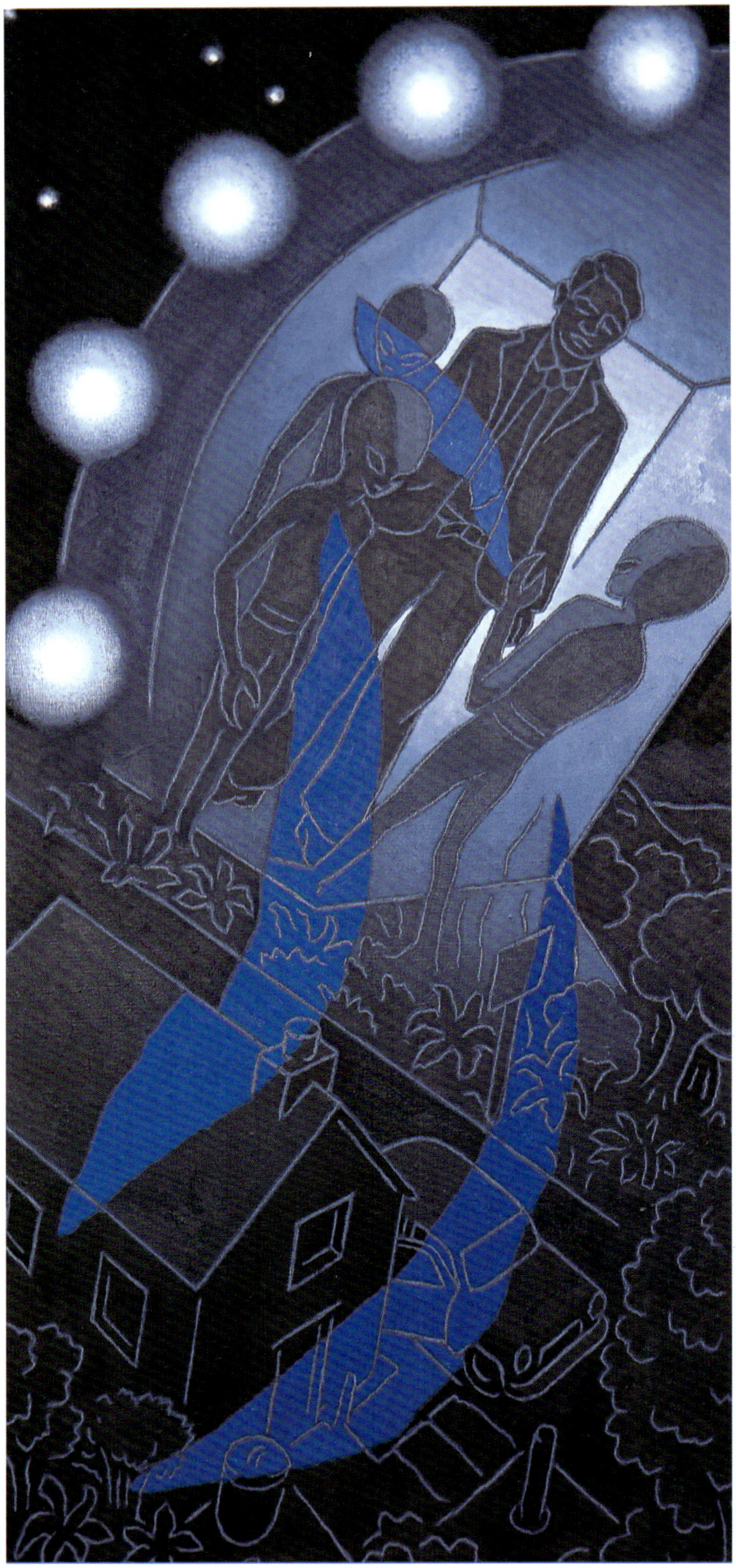

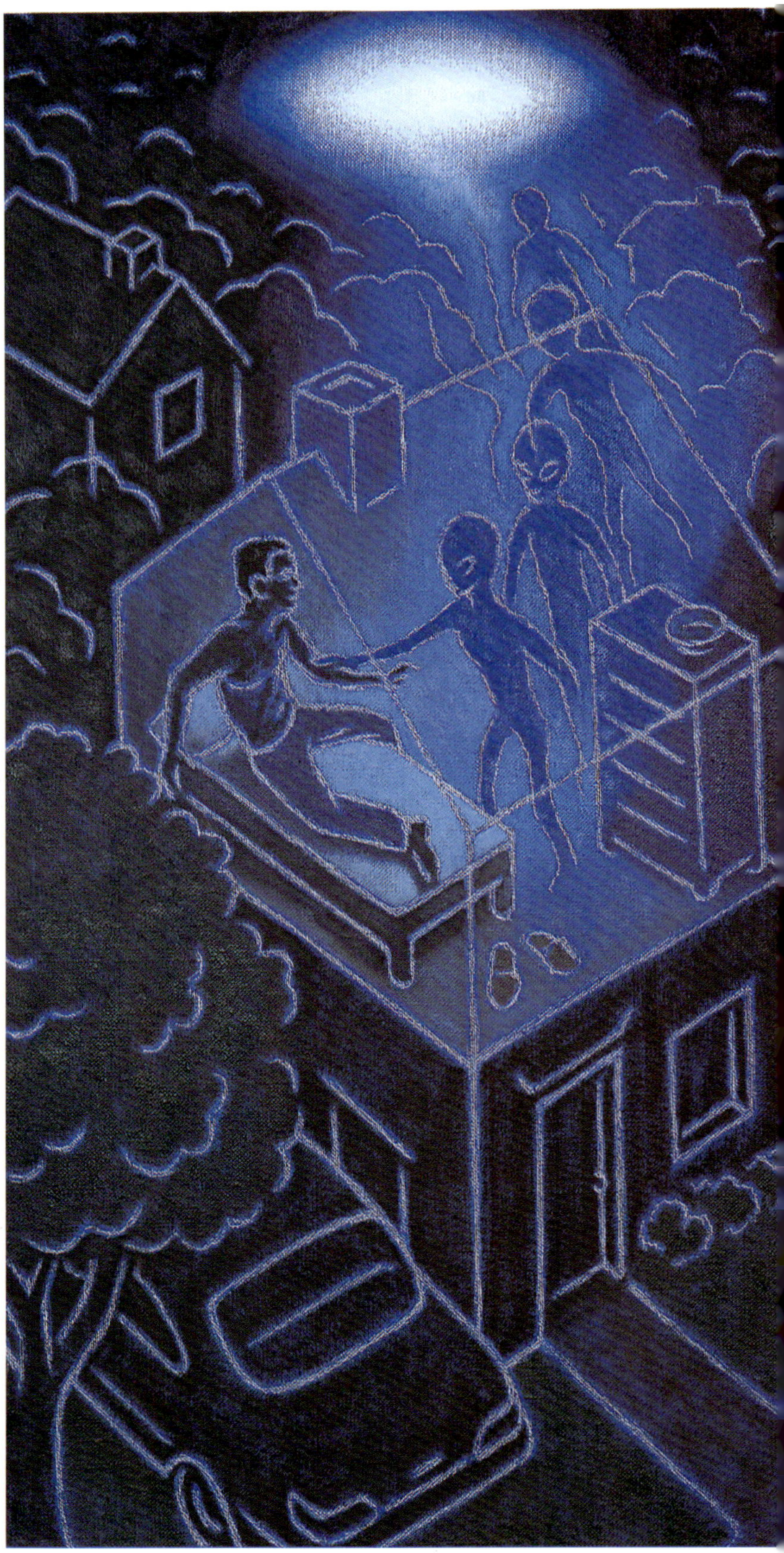

THE ALIEN IMAGE

MICHAEL BUHLER

In 1996 the biggest selling poster in the United States featured the sinister image of an extraterrestrial being, commonly known as the 'Grey'. This creature's name derives from its typical appearance: a mushroom-coloured complexion startlingly contrasted with distinctive, black, almond-shaped eyes and a mean slit-like mouth. Its disproportionately large, pear-shaped head is supported by an emaciated body. In the United States the image of this grim creature has become synonymous with alien abduction stories.

Whether or not UFOs and aliens really exist, they are a growth industry. Numerous new glossy magazines are full of advertisements promoting the Greys on T-shirts, caps, badges, key-rings, skateboards and other merchandise. Wherever you look, you are aware of the dark, malevolent eyes of the watchful Greys, who have successfully wiped out other representations of alien species and are universally accepted as the alien archetype.

However, the Greys have not always been our sole extraterrestrial visitors. In the 1950s and 1960s there were many 'sightings' of creatures emerging from stationary flying saucers. Hundreds of reports of these sorts of encounters from all parts of the world were collected together in 1966 for the publication *The Humanoids*. Descriptions of what these aliens looked like were extremely varied. There were enormous differences in the size, shape, facial features, colouring, clothing (or lack of it) and the manner in which they moved in the Earth's atmosphere. Some extraterrestrials seemed to be comfortable with our gravity and atmosphere, walking around robotically, or floating over the ground like wraiths, able to breathe our impure air. Others were less at home, wearing space-suits, helmets and breathing apparatus.

Despite this diversity of appearance, there was a surprising and consistent similarity in the behaviour of all these alien creatures when in contact with humans. In most accounts, when they were spotted outside their saucers they would hastily get back into their craft and take off, leaving little or no trace of their presence. It was difficult, however, to reconcile the variety and behaviour of these visitors with a unifying extraterrestrial hypothesis: there seemed to be too many different species, too many sightings, and very few measurable facts.

Researchers looking for harder evidence sought other explanations. Analysing the reports given by witnesses, they noted that some of these creatures closely resembled the imps, fairies, dwarves and elves of folklore, and suggested that we might be re-interpreting in terms of our own technological preoccupations a phenomenon that has been with us for centuries.

In his book of 1969, *Passport to Magonia,* Jacques Vallee was the first writer fully to develop this particular idea. Speculation that these creatures may not have come from outer space at all increased: they might always have shared the planet with us, existing in some other dimension. There was no evidence to support this, however, and other European ufologists explored the possibility that the whole phenomenon could be explained in psycho-social terms, basing their theories on Jung's *Flying Saucers – A Modern Myth of Things Seen in the Sky*, written in 1959. They made comparisons with traditional beliefs such as witchcraft and studied the many flourishing UFO cults.

Nobody was any closer to a satisfactory explanation, and in the United States, these new approaches to the problems were hardly noticed. Despite scepticism, the UFO 'community' continued to think in terms of space visitors and ignored the descriptions of creatures who could be hypothesised as manifestations from fairyland rather than visitors from Planet X. This self-censorship encouraged the support of an already emerging acceptable alien archetype, fuelled by an appetite for television programmes ranging from factual documentaries to science fiction fantasies.

Perhaps the Greys began their invasion in 1947, in the most famous 'extraterrestrial' incident, which took place near the town of Roswell, New Mexico. On the night of 2 July, during a heavy thunder storm, William (Mac) Brazel, a rancher, heard a loud explosion above the noise of the storm. The following day, while checking his sheep, he and his companions found a wreckage that seemed to be the remains of a craft, made from a substance like metal foil and spread out over a few hundred square yards. Some pieces of the debris carried strange symbols and texts. Their first thought was that something had crashed on the site and that it was probably an Air Force plane.

On 7 July Mac Brazel drove 70 miles to Roswell, the nearest Air Force Base, and reported his find. An immediate investigation was launched. The debris was collected and the area cordoned off. A startling press announcement claimed there was evidence to suggest that the wreckage of a flying disc had been recovered. The same day a civil engineer came across another find several miles away from the first: the main

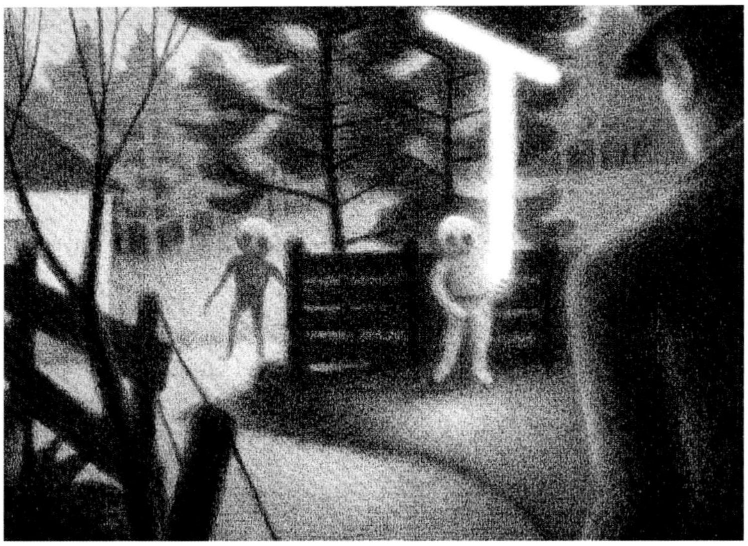

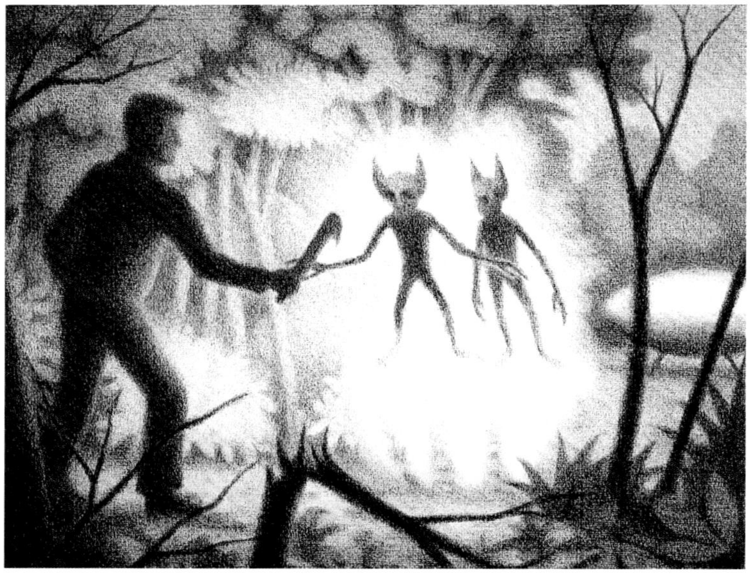

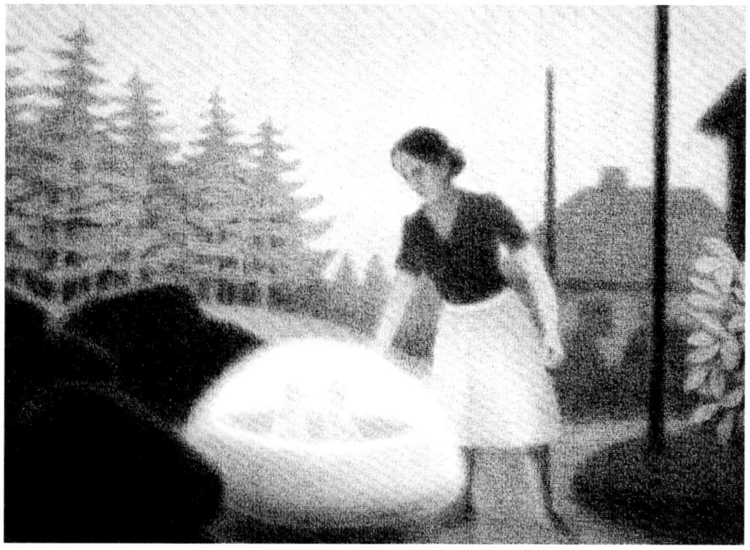

Michael Buhler, FROM ABOVE: Pine Township, USA*;* Vincenza, Italy*;* Rauma, Finland, *all conte on paper, 38.1 x 53.3cm; OPPOSITE:* Intruders, *oil on canvas, 101.6 x 81.2cm*

body of the craft, containing a number of small humanoid corpses with large heads, dressed in grey one-piece suits.

Following this new information, the Air Force called a further press conference to announce that the earlier statement was a mistake and that the debris was that of a weather balloon that had been mis-identified. Following the retraction of the claim of a UFO crash, the press lost interest and the wreckage was then secretly flown to Wright Paterson Air Force Base for further investigation.

To this day the events around Roswell remain a mystery but there is no doubt that something unusual was recovered in the desert. The current position held by the Air Force is that the balloon was part of a top-secret project to monitor Soviet nuclear testing and was definitely not a UFO. However, many official authorities dispute this.

Confused and contradictory reports have persisted since the original event, and those wishing to document or provide accounts of the incident have had to select from a large number of different versions of events and conspiracy theories. This one incident has been the subject of four books, countless articles, several television documentaries, a feature film and a best-selling video claiming to show the autopsy of a dead alien retrieved from the wreckage of the crash. The media has thrived on Roswell, which now has a tourist industry and its own UFO museum.

Up to this point in alien history, the Greys had hardly featured in any accounts of sightings and neither were there any records of abduction cases. This began to change when the best seller, The *Interrupted Journey* by John Fuller, was published in 1966. It described what was to become a typical account of an abduction and included drawings and descriptions of the abductors, whose images were close, but not identical, to the archetypal Grey.

The couple featured in this account, Betty and Barney Hill, had allegedly been intercepted whilst on a nocturnal car journey, but had little or no conscious memory of the event. They claimed to have spotted a star-like light moving across the sky and after this sighting they arrived home two full hours later than expected. They were left with troublesome nightmares and other symptoms of distress, which forced them eventually to seek help.

Two years after the encounter, they agreed to hypnotic regression in an attempt to recover the full story of that night. This technique is routinely used to help people recall events, but there is no guarantee that these 'memories' are accurate representations of real occurrences. The Hills' story unfolded to reveal that after they saw the travelling light, their car had been forcefully stopped and transported onto a space craft. The occupants were about six feet tall with large heads and almond-shaped eyes. They proceeded to conduct full medical examinations on both abductees, taking skin and blood samples. The creatures even showed Betty a star map of their

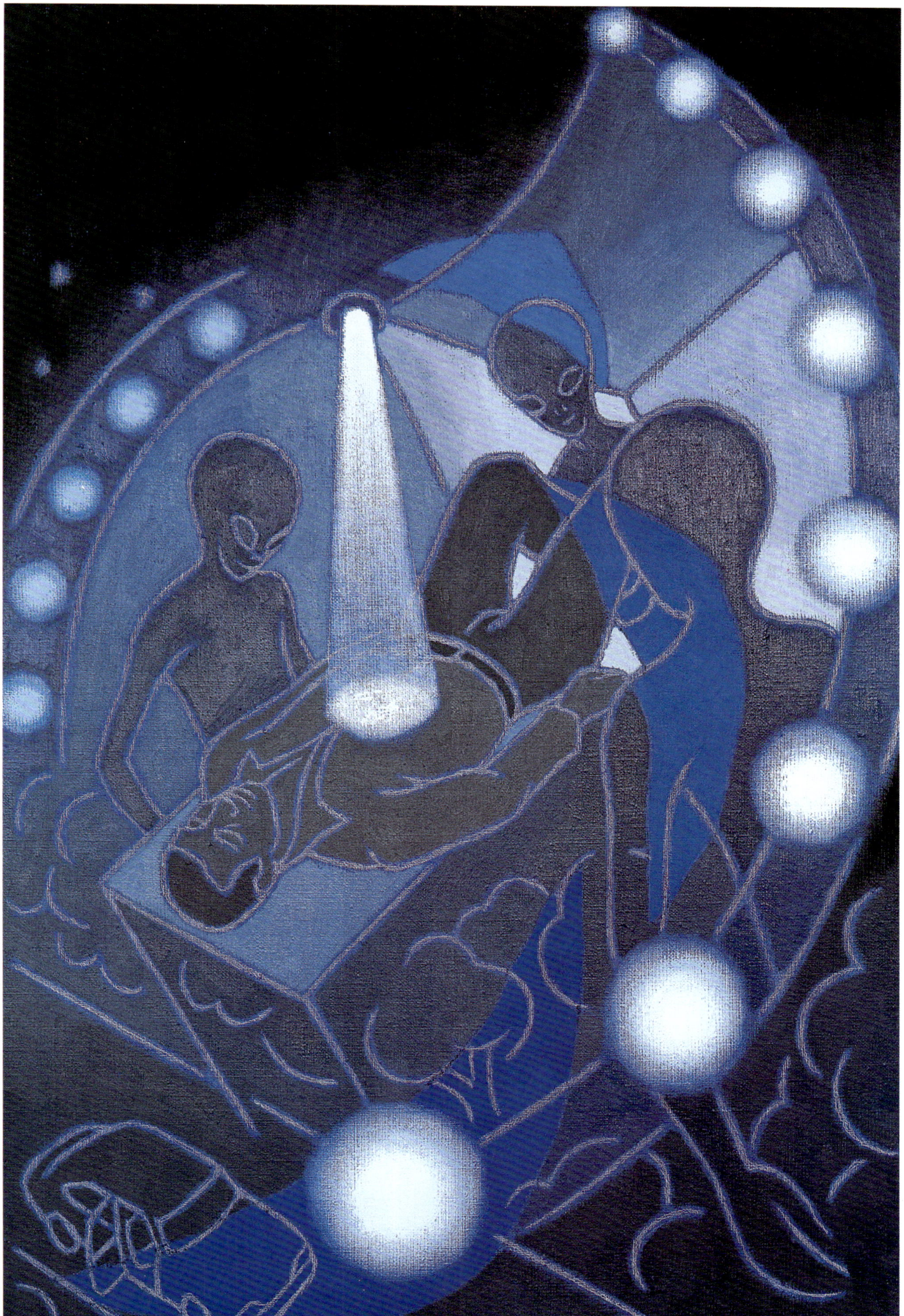

space-trading routes before the couple were told to forget everything they had seen and allowed to drive home.

By the end of the 1970s, two significant publications on abduction had been released. One told the Hills's story, and the other was a book on Travis Walton, another 'abductee'. Travis Walton was one of seven members of a wood-cutting gang who were travelling home through the forest on 5 November 1975. Driving over the top of a small hill, they were stunned to find a disc-shaped UFO that gave out a golden glow. Walton ran towards the craft, which fired a blue ray, knocking him back, and the driver of the truck sped away from the site before any of the others had time to go to Travis's aid. A short time later, the six remaining men returned but there was no sign either of the craft or of Travis. He had vanished.

Despite a police investigation, Travis stayed missing for another five days, reappearing on 10 November in a state of confusion. He said that he had been abducted by short grey aliens with large heads and big eyes. As the first recorded event witnessed by a number of people, the Travis Walton abduction received massive media attention. This story has subsequently been made less verifiable by evidence suggesting that the co-workers could have colluded with Travis in a hoax.

In 1981, the American artist and ufologist Budd Hopkins published his first book, *Missing Time*, which focussed entirely on abductions. This work included drawings by the abductees of Greys and relied on hypnotic regression for recall of events. Hopkins's surprising conclusion was that these people had not just been abducted once but were actually specially chosen human specimens whose first abductions had begun in childhood.

As the 1980s progressed, the Grey came to be regarded as the villain responsible for all abductions, and this was reinforced when, in 1987, the psycho-thriller writer, Whitley Strieber, wrote his best seller, *Communion*. A 'true account' of his own recent abductions revealed under hypnosis therapy by Budd Hopkins, the book was aggressively marketed. Both the cover and the street posters featured the hairless head of a sinister alien with deep, black, staring eyes, malevolently watching for the next victim.

Hopkins published his second book, *Intruders*, soon afterwards. Again, it featured the Grey, but more importantly, it proposed a frightening explanation for the abduction phenomenon: abductees were the unwitting participants in a programme to cross-breed humans with Greys. Hopkins found plenty of people who thought they might be involved in this bizarre project and who would offer their own accounts and impressions for publication, including drawings of the Greys themselves and their hybrid, waif-like children. Despite mounting objections, such as the technical impossibility of crossing alien and human DNA, or inconsistent portrayals of the sophistication of alien science, Hopkins remained constantly in demand for interviews, articles and lectures.

In the 1990s, abductions dominate the American UFO scene, and books and popular television programmes such as 'The X-Files' and 'Dark Skies', which regularly feature the Greys, continue to promote the endorsed portrait of alien visitors.

Comparing the accounts of a number of 'abductees' in his book *Secret Lives* of 1992, Dr David Jacobs concluded that there was a consistent procedural pattern running through nearly all the cases: the Greys were taking sperm and ova from humans and using them for genetic experiments. People were not being purposefully harmed but treated like laboratory animals, many suffering mental and physical trauma.

In 1993, 'aliens' bearing no relationship to the archetypal Grey were encountered in Australia. These sightings involved multiple witnesses, who are willing to be identified, and perhaps most importantly, were unknown to each other before the event.

Three cars, travelling very late at night, stopped to observe a large UFO that had landed near the road. The creatures who took the six people involved onto the UFO were about seven feet tall with long bald heads and huge, round, red eyes. The witnesses underwent a series of medical examinations and had no immediate conscious recall of their experience. The case is still under investigation, but the claim that these people experienced something entirely real and physical has gained the support of many ufologists.

The current situation for serious ufologists is that the public everywhere is now far more aware of the possibility of alien abduction than ever before, and there is therefore much confusion between what might be the real and fictitious elements of the different accounts. Some researchers are convinced that there is more to the phenomenon than can be explained in terms of psycho-social myth, and are encouraged by the new evidence of different images, however strange and disturbing. But it is the media's obsession with marketing a universally recognisable image of an alien archetype that has championed the sinister face of the Grey. This threatening species haunts the 1990s, overshadowing any other potential, and perhaps, more realistic image.